Mathematical Misconceptions

Mathematical Misconceptions

A Guide for Primary Teachers

Edited by
Anne D. Cockburn and Graham Littler

Los Angeles | London | New Delhi
Singapore | Washington DC

First published 2008

Reprinted 2010

SAGE Publications Ltd
1 Oliver's Yard
55 City Road
London EC1Y 1SP

SAGE Publications Inc.
2455 Teller Road
Thousand Oaks, California 91320

SAGE Publications India Pvt Ltd
B 1/I 1 Mohan Cooperative Industrial Area
Mathura Road
New Delhi 110 044

SAGE Publications Asia-Pacific Pte Ltd
33 Pekin Street #02-01
Far East Square
Singapore 048763

Library of Congress Control Number 2008924327

British Library Cataloguing in Publication data

A catalogue record for this book is available from the British Library

ISBN 978-1-84787-440-5 (hbk)
ISBN 978-1-84787-441-2 (pbk)

Typeset by C&M Digitals (P) Ltd, Chennai, India
Printed in Great Britain by CPI Antony Rowe, Chippenham, Wiltshire
Printed on paper from sustainable resources

Contents

List of figures

List of pictures

List of tables

Introduction

Pause for thought

Do you know anyone who feels neutral about maths?

We suspect not. Indeed, we would go so far as to suggest that the majority of people either love it or hate it. Whether we like it or not, however, we are surrounded by numbers and most find it hard not to engage with them several times a day when we phone a friend, search for a room number, buy a newspaper and so on.

Many factors influence how we feel about the subject. Some we will be aware of, but others may be less obvious. Many will be able to conjure up a primary teacher who instilled in them a great love or loathing for the subject. Some will remember bafflement and a feeling of isolation followed by a lack of confidence while tackling a problem, while others will recall the sheer exhilaration of encountering a challenging mathematical investigation. You may think back to occasions when you tried to unlock the key to find a pattern or an easy route to a solution: whether you were generally successful in these quests or not will almost certainly have influenced how you feel about mathematics today.

This book is about how people – in this case generally primary school children – think about mathematics: how they try to make sense of it and reasons for some of their successes and misunderstandings. We have focused specifically on number as this plays a crucial role in the earliest years of schooling and often has a major influence on how we feel about mathematics in later life.

This book is also about broadening our understanding of numbers and how we use them. Some of the chapters – Chapters 0 and 1 – assume very little mathematical knowledge, making them accessible to even the most reluctant readers. Others – most notably Chapter 7 – were specifically written for the more mathematically minded but, if taken slowly, provide a wealth of information useful to us all.

This is a book for anyone working – or about to work – with 3–11-year-olds. Most readers are likely to be prospective or experienced teachers working in schools, but some will be teacher educators, parents, teaching assistants or others interested in children's thinking. A few may even pick it up off the bookshelf to see whether it might enhance their mathematical knowledge and understanding. As a team of teachers, teacher educators, mathematicians and psychologists we hope that you will find something of interest. Indeed, our endeavour is that, whatever your knowledge of classrooms and mathematics, this book will increase your understanding of number, how we learn about it and how to enhance experience of engaging children in what some see as the mysteries and others see as the delights of the subject.

A biographical note

This book was born in Italy when Anne – a psychologist, mediocre mathematician and primary teacher educator – first met Carlo – a professor in advanced mathematics at the University of Parma – and began discussing why 'zero' is plural in English: we say 'one dog' but 'zero dogs', for example. This led us on to the idiosyncrasies of different languages and the wide scope for misunderstandings when children first encounter mathematics – which itself is a complex and challenging language. In a surprisingly short time the idea for an international project on mathematical misconceptions emerged and colleagues from Israel and the Czech Republic were invited to join the team. A bid for funding was submitted to the British Academy, and we began work on 1 August 2006. The team then expanded to include teachers from across the primary age range in each of the four participating countries, making for a highly experienced and diverse group of experts. Much stimulating discussion and debate ensued as we grappled to understand our different perspectives. We are now all much better informed, and this book reflects our very differing areas of expertise and interests. Each chapter has one, two or even three lead authors, but we have all read, commented on and contributed to all of them, making them – we hope – richer than if written independently and in isolation (see below). Having said that, as may be apparent from the following alphabetical listing, all the contributors are experienced writers in their own right:

Anne Cockburn	School of Education and Lifelong Learning, University of East Anglia, Norwich
Sara Hershkovitz	Center for Educational Technology, Tel Aviv
Darina Jirotková	Department of Mathematics Education, Charles University, Prague
Graham Littler	Visiting Professor of Mathematics Education, Charles University, Prague
Carlo Marchini	Professor of Complementary Mathematics, University of Parma
Paul Parslow-Williams	School of Education and Lifelong Learning, University of East Anglia, Norwich
Fiona Thangata	School of Education and Lifelong Learning, University of East Anglia, Norwich
Dina Tirosh	Professor in Mathematics Education, University of Tel Aviv
Pessia Tsamir	Department of Mathematics Education, University of Tel Aviv
Paola Vighi	Professor of Mathematics, University of Parma

The book is arranged in three parts. The first (Chapters 0, 1 and 2) makes few assumptions about readers' background and expertise and focuses on largely mathematical matters. The second (Chapters 3–6) adopts a more psychological perspective, while the third (Chapter 7) provides a more in-depth examination of the mathematics taught in primary school classrooms. Chapters can be read in any order but were designed in a specific sequence which favours the more conventional approach of starting at the beginning. For example, from time to time, we refer to various tasks which reappear on several occasions throughout the book. At first sight some of the chapters may not look particularly relevant to you because, for example, you teach a different age than that discussed initially or the mathematics looks too simple. Please take a minute, however, to read three or four pages, for one of the major findings of our work was that a broader appreciation of the number work done across the primary age range reduces the likelihood of mathematical misconceptions arising and enhances both the quality and enjoyment of the educational experience.

Three important starting points

As the title suggests, much of this book discusses mathematical misconceptions and, in particular, how they arise and how they might best be managed. It is important to emphasise right from the start that we do not view misconceptions as a 'bad thing'. On the contrary, they often reveal much about children's thinking and how they acquire – or not, as the case may be – an understanding of complex mathematical concepts. Indeed, they can often be used as useful discussion points in lessons as they can highlight pupils' ingenious – albeit possibly incorrect – ways of tackling a problem.

We also wish to stress that we all have a tremendous regard for primary teachers and the way you work so tirelessly with numerous children and a myriad of demands day in, day out. Having worked in schools, we can appreciate where your enthusiasm might come from, but your stamina and ability to multi-task non-stop are another matter! We wish to reassure you, therefore, that this book is not about pointing the finger at anyone but rather about exploring the many ways in which misconceptions may arise.

This brings us to our final starting point. We see young children starting school as highly knowledgeable, creative and imaginative individuals who have had a wealth of experience with numbers before they ever enter the school gate. Below are some cameos of life in nursery classrooms which illustrate these very points.

Cameo 1. The library in the nursery class was displayed as shown below. Sam announced, 'I want 3 as I'm 3', and promptly removed the book which was in the '3' slot – *Animal Treasury* – and replaced it with the book he had just been looking at.

'I want 3 as I'm 3'

Cameo 2. Outside, four 3- and 4-year-olds were working with their teacher, who took on the role of the queen bee. Each in turn was given a choice of four or five hexagonal cards. These had a numeral (between 1 and 9) on one side and the corresponding number of dots on the other. The children's task was to select a card, search for the 'flower' with the corresponding number on it and collect the amount of honey (i.e. bricks) specified on their card and the flower.

The honey collecting task

Joey picked out a card with the numeral '9' on it. He duly rushed off to the '9' flower, withdrew nine bricks from the bucket and returned to the teacher. He repeated a similar operation to collect '5 bits of honey'. Spontaneously he merged his two collections of bricks and successfully counted the lot.

Cameo 3. A nursery teacher stuck the following numbers in this order on the whiteboard: 3, 1, 2. A child immediately announced, 'Number 1 should go first, then number 2, then number 3'.

Cameo 4. Jack (aged 3) wrote '3', spontaneously counted three building bricks and then pointed to a picture in the play area saying, 'That's 3 little pigs. I'm 3'.

The 3 little pigs

Cameo 5. Patsy, aged 3, completed her 'bus stop task' by writing '4' on the stop. Without any comment from the teacher she then wrote another '4' followed by the numbers 1 to 4 twice. She counted all the numbers she'd written and announced that there were 10.

Patsy's bus stop task

Cameo 6 comes from Italy. The children were asked to draw or write a story about the numbers 0, 1 and 10. Gina's thought-provoking response has been reproduced below. For the non-Italian speakers among us 1 is saying to 0, 'You are only nothing'.

'You are only nothing'

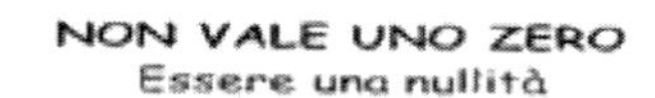

Acknowledgements

Our especial thanks go to the British Academy (award no LRG-42447) without whom this work would not have been possible.

We are also extremely indebted to all the participating primary teachers and pupils who gave so generously of their wisdom and time.

Thanks, too, to Fiona Thangata who began work on the project before going off to the sunny climes of Botswana.

We are also most appreciative of the work of Helen Fairlie and her team at Sage for having faith in our project and always being willing to provide quick and invaluable feedback.

Finally, the team are tremendously grateful to their families and, in particular, to those who produced grandchildren during the period of the project: they reminded us what it was all about!

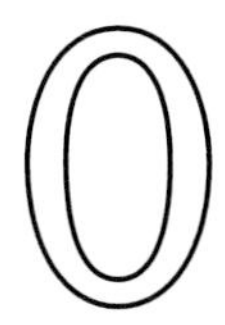

Zero: understanding an apparently paradoxical number

Anne D. Cockburn and Paul Parslow-Williams

Challenge

How would you describe zero?

Do you think of 'zero' as a four-letter word that comes towards the end of the dictionary? Or do you think of it in terms of nothing at all? Or do you simply wish that your bank statement included more of them, especially after a '9'! Whatever your response it is likely that you – along with the vast majority of people – will realise that zero is a kind of paradox: sometimes it is used to signify that you have, for example, 'nothing' to add, as in $4 + 0 = \square$. And yet, if simply preceded by another digit, it can imply that you have a significant amount to add, as in $4 + 90 = \square$.

As will become more apparent as you read this book, the conundrum many people experience is not really knowing what zero means. Every day we see '0' in a wide variety of settings – in a telephone number, on birthday cards, as part of a house number, when weighing items and so on – but rarely do we stop to think about it. We recently heard the following conversation with Eva – a bubbly 3-year-old in a nursery class – and her teacher, Mrs Clarence. Eva has an older sister and a younger brother.

Anne: Do you know how old your sister is?
Eva: 8
Anne: And how old are you?
Eva: 3
Anne: And how old is your brother?
Eva: Zero ... I don't know ... one
Teacher: [whispering to A] Not quite one.

Clearly Eva's brother was not 'nothing' otherwise he would not exist but nor, indeed, was he 'one', so what was he! As will be discussed in more detail in Chapter 7, in this context zero belongs to an infinite set of numbers, including 0.75 years old, which might otherwise be translated into 9 months and the age of Eva's brother. The words we use for counting objects are called *cardinal* – or *natural* – *numbers,* and it is these numbers that adults tend focus on when introducing children to early mathematics. In this chapter we want to explore some of the challenges of working with zero and the misconceptions which can arise through early misunderstandings.

First experiences of zero

Pause for thought

When did you last use the word 'zero' in everyday conversation?

If you stop to think about it, most of us simply do not include 'zero' as part of our everyday language. Do you ever use it at home or as you go about your daily life? Have you ever come across a nursery rhyme, song, story or poem which includes 'zero'? You may be familiar with some which include the concept of zero – such as 'Ten green bottles' or 'Five little frogs' – but we suspect you have never encountered the specific word in any of them. Thus it usually falls to the early years teacher to introduce the appropriate mathematical language of zero together with an explanation of what it means. Typically this will involve a discussion which includes words such as, 'nothing', 'none left' and 'empty'. This is definitely one aspect of zero but, as described at the beginning of the chapter, it is not the only one. Our observations, however, suggest that it is very much the dominant one experienced by children which, as a result, can create problems for them when they encounter other definitions. It may be helpful to think of the process in terms of an analogy: imagine a sapling growing at the top of a windswept hill. Once planted, it experiences a wide range of winds and weathers (similar to all the zeros children encounter in their everyday lives) but, over the months and years, it may begin to bend in one direction if the prevailing wind (for which read: an emphasis on 'nothing' from parents and poems etc. to the almost complete exclusion of other facets of zero) is particularly persistent and strong. If no remedial action is taken the tree may never recover and thus, rather than realising its full potential as a magnificent oak stretching skywards, it may become bent and twisted out of shape. An observant gardener, however, might spot the potential for damage and provide a stake for the tree in its youth which, without undue pressure, encourages it to develop into a strong and healthy specimen. So, too, with the formation of multifaceted concepts: if you focus on and encourage only one side or dimension – such as the idea that zero means nothing – at the expense of another it significantly increases the likelihood that a child will develop a distorted view of the concept and fail to realise their full mathematical potential. Indeed, this might occur to such an extent that it becomes increasingly difficult for them to see the concept – in this case, zero – in its full glory, including its role in 10 and many of the numbers beyond. In passing, such alliances sometimes occur with other concepts. With subtraction, for example, children who have spent a lot of time 'taking away' – as opposed to comparing – can find it difficult to appreciate that the minus sign has several different, but related, interpretations. The temperatures –4°C and –9°C are relatively easy to compare, for example, but are literally impossible to take away from one another. The concept of subtraction is discussed in more detail in Chapter 3.

So how can we provide a stake for our vulnerable saplings?

Representing zero in the early years of schooling

The concept of zero is clearly not an easy one to acquire and it appears that the way in which we say numbers in everyday life may not help. For example when talking about high-speed trains you generally hear 'one, two, five' rather 'one hundred and twenty-five', and we have never heard a telephone number described as 'eight hundred and sixty-five thousand, five hundred and twenty' as opposed to 'eight, six, double five, two, oh'. So how do experienced early years teachers tackle the issue of zero's multiple personalities? How do they, for example, portray it as a number representing 'nothing' (i.e. as associated with the empty set) as well as a digit playing a fundamental role in many of the numerals – such as 10, 307 and 50679 – greater than 9? This apparent paradox engaged the teachers in the project group in a lively discussion of what they currently did and what they might do in the future:

Sandy: I think it is important that zero is displayed alongside the other numbers in the classroom in a wide variety of ways.

Annabelle: Yes, I agree. I have all kinds of numbers around the classroom: phone numbers (pretend, of course!), house numbers, birthday and date charts, number lines ...

Sandy: Someone suggested that I have a zero to ten 'washing line' with shirts on it labelled '0' to '10' but I think that's confusing as should one count the shirts or focus on the numbers? Instead I hang string bags from a number line. The first one – that is, hanging from zero – is empty, the next one contains one cube, the next two and so on.

0 in the bag

Annabelle: I really like that idea! At the moment I have a vertical number line with '0' at floor level rising up to '10', but I am now wondering if that might create problems when we come to negative numbers later.

Ruth: I worry about showing empty palms to represent zero as they could be seen as showing 10 fingers. I much prefer showing my fists for zero and then raising one finger for one, two for two and so on.

A fist of zero

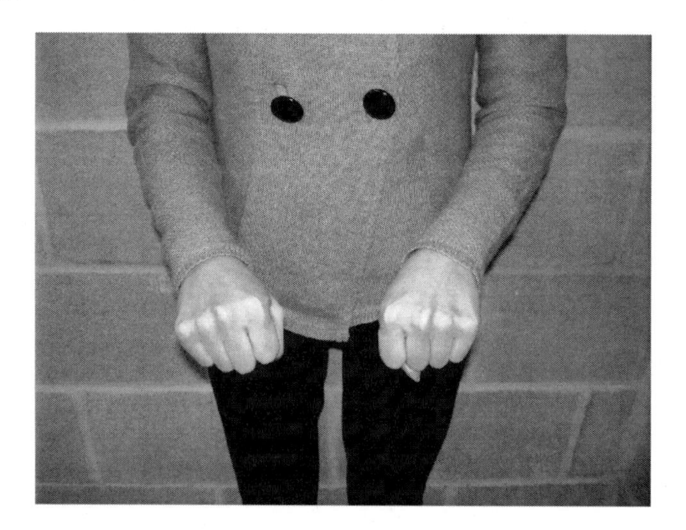

Annabelle: In the past we always counted up from 1 to 10 but down from 10 to zero. I'm now going to try counting up from zero. It may be unconventional but it is worth a try!

Ruth: I quite like the standpoint of the same approach for each digit. I have always given my children a card with a number on it and asked them to get the same number of cubes as shown on the card. Now I could include a card with '0' on it and make sure that there is a table with nothing on it in the room and ask the children to show me zero and they've actually got to show me something. Of course they might not point to the table but if they pointed to a space on the carpet or showed their fists that would be fine.

Sandy: I think it's crucial that we break down the 10 barrier in their thinking and that our teaching programme isn't limited to that 0 to 10 idea, but if you give teachers a curriculum [here she is referring to various government documents such as the *Primary Strategy* and the *National Numeracy Strategy*] and a statement connected to it that's what you are going to get. I tend to use the teens numbers really early on but I don't talk about unpicking them and writing them. I think we need to do the bundling of 10 and 1 for 11 earlier on. I'm not sure how you explain the one and the zero. You can't say to little children that you don't want too many marks but you need to re-educate them and make them see that it's almost like a new beginning. I think you need to spend a lot of time 0 to 6. I don't think 0 to 5 is particularly useful: you can do so much more with 0 to 6 as there are many more number bonds you can discuss.

In essence, zero tends to be the first 'mathematical' – rather than everyday – number children encounter. As such, it is often treated as a special – and perhaps rather challenging – entity. If, however, it becomes more of a feature of daily classroom life, children are more likely to develop a fuller understanding of what it means. Thus, for example, when distributing five crayons between two children you might give four to Helen and one to Jack, or three to Jack and two to Helen, or zero to Helen and five to Jack and so on and observe the lively debate which would almost certainly follow! With older children discussions on place value could include just as many examples with '0' as '9' or '4' or any other digit. Comparisons between the numbers of tens in 324 and 304 could be made, and the validity of '020' could be considered in a variety of contexts including London telephone numbers. Such activities encourage children to see zero as a number of importance rather than a nonentity. For a further discussion on this issue, see Chapter 6.

Below we will discuss some of the confusion zero appears to create but, before we do, we think it might be helpful to digress ...

What is place value?

> **Challenge**
>
> Take a couple of minutes and jot down what the term 'place value' means to you.

The 'place' in place value is certainly key but it is not just about position: the five on the left-hand side of this page is not worth more than the one on the right-hand side for example, even though it might be higher up on the page.

5	
	5

The place a digit takes relative to another is also crucial if they are within a certain distance of one another – but how big might that distance be? How many numbers do you see in the line below, for example?

3 1   4 2   4 6   6   4   1 2 3 4

As I (i.e. Anne) typed them I was thinking '31, 42, 46, 64' and then '1, 2, 3, 4', but there is no reason to assume that this 'more correct' than your response, especially given that, if you pair the digits from left to right (or vice versa if you prefer), the distance within each pair – regardless of the physical size of the number – was created by pressing the space bar on my computer keyboard twice. It may well be that, as an adult with experience of place value, you saw the number combinations as we did, but there is no obvious reason why a child should do so. Indeed, any response between one and twelve is reasonable although, we suspect, given the distribution of numbers, some answers are more likely than others.

Moving on to consider a further aspect of place value, when presented with the two-digit number '55', the 5 digit to the left is considered to have more value – in terms of quantity – than the one on the right.

Challenge

If we change one of the digits in 55 – but do not tell you which one – what can you tell us about the newly formed number?

Assuming that no zeros were involved, you can say with confidence that the value of the left-hand digit is still higher than the one to its right. Thus, for example, the new number may be 15 but the left hand digit is worth twice the amount of the one to its right. Had the transformation been to 59 the left hand digit would have been worth considerably more than the 9 to its right. In other words, to appreciate place value one requires a knowledge of the *value* of a digit and its *place* relative to any other digits in the number under consideration. This is discussed further in Chapter 2.

This, to an adult, may seem obvious but, if you stop to think about it, place value is both a complex and surprisingly abstract concept to present to a young child who is only a few years out of nappies and is at the very early stages of beginning to read and write.

Some historical highlights

Ifrah (1998) observed that in Egyptian hieroglyphics there was no need for zero as numbers such as 2004 could be represented by two lotus flowers (for thousands) and four vertical lines (for units) or, in terms more familiar to us,

$$1000 + 1000 + 1 + 1 + 1 + 1 = 2004$$

This is similar to the Roman idea of, for example, using 'C' to represent 100 and 'X' for 10, making CCXXX equivalent to $100 + 100 + 10 + 10 + 10 = 230$. Interestingly, in one of these systems the order of the symbols is irrelevant: which?

Pause for thought

Take a moment to think whether you would have the patience and ability to represent the number 999999999 using Egyptian hieroglyphics. How many items would it involve? (You should assume that there is a different symbol for each decimal place and that, as above, each symbol represents one of something, e.g. one lotus flower = 1000 and one vertical line = 1 unit)

Kaplan (1999: 4) makes what to many nowadays might seem like an obvious point:

> we count ... by giving different number-names and symbols to different sized heaps of things: one, two, three ... If you insist on a wholly new name and symbol for every new size, you'll eventually wear out your ingenuity and your memory as well.

It is thought that zero was introduced in India in the sixth century and that originally it was represented by a dot rather than a small circle (Flegg, 1989). Its birth removed the need for numerous symbols to be used. Thus, in the Hindu-Arabic system with which we are all familiar, for example, we can create any number – large or small – using a combination of only ten digits: 1 to 9 and, most crucially, 0. Returning to the question above, in the Hindu-Arabic system the number 999999999 is, in effect, shorthand for

$$(9 \times 100000000) + (9 \times 10000000) + (9 \times 1000000) + (9 \times 100000) + (9 \times 10000) + (9 \times 1000) + (9 \times 100) + (9 \times 10) + 9,$$

which is long but not quite as cumbersome as the 81 items you would need to add if you were using Egyptian hieroglyphics.

Developing an understanding of numbers over 9

Our discussions with teachers suggest that children as old as 8 and 9 years frequently demonstrate misconceptions when counting. For example, a year 4 teacher in the UK described how 'When they count backwards they will go from 201 to 199 missing out the 200'. She went on to explain that she thinks 'They see the pattern as 9-1-9'.

Sandy – a reception teacher who suggested beginning with the numbers 0 to 6 earlier in the chapter – explained the problem as she sees it:

> In the *Early Learning Goals* it recommends that 4- and 5-year-olds learn the numbers up to 10. The result, in my view, is that children almost see ten written as '10' as a single 'digit', without understanding. All this focus on numbers up to and including 10 but no further can create problems. The 'ten-ness' of ten is not explored and, indeed, I don't think it can be without comparison with 11 or 12 etc. Would it be more helpful to begin with 0–9 and then move on to 10 and above when they are ready to understand early concepts of place value?

Pause for thought

The Italians have numerous words for pasta, but imagine you are going to Italy for the first time and the only word you know for pasta is 'spaghetti'. You have been asked to make a meal including pasta but you only have a very vague idea as to what it actually is. When you try to buy some, the shop has run out of spaghetti. The shopkeeper does not speak English and you do not speak Italian, how might you discover what pasta is?

Stopping to think about it, most of us find it far easier to develop an understanding of a rule if we are introduced to several examples of it and counter-examples which do not follow the rule. Thus, with the challenge above, it would be helpful if you were shown spaghetti, linguine and ravioli as examples of pasta and asked to contrast them with potatoes and polenta so that you could appreciate what constitutes pasta and hence decide which to buy. So too with numbers, if children are introduced to several two-digit numbers including 10 at the same time and then asked to compare and contrast them with single-digit numbers, then we suspect they will be far more likely to understand the early principles of place value and the significance of zero as a place holder.

Graham – one of the authors of Chapter 6 – recounts the following experience:

> When I was teaching this age group, 5–7 years, I used home-made equipment of cocktail sticks and bundles of 10 cocktail sticks stuck together – ensuring they all had the sharp points cut off. I introduced the numbers 0 to 9 and we did calculations of addition and subtraction using these. Then I always went to addition greater than 10 because the pupils put their sticks on a card which had two columns, sticks and bundles and if they had to add 6 + 7 then they would put six sticks followed by seven sticks in the 'stick' column. They would then separate 10 single sticks and exchange this for a bundle and finally write down how many objects they had in each column, so in this case it would be 1 bundle and 3 sticks. I then went back to tasks such as 4 sweets and 6 sweets and found that the pupils did not forget to put the 0 in the sticks column when they had changed their 10 sticks to 1 bundle whereas before they had forgotten to put the 0 in.

A very common problem

Donna (year 4) reported that her pupils encountered difficulties with numbers greater than 1000, noting that

> the children often want to say 'one thousand two hundred and fifty', but they write it as 1000250.

She went on to explain why she thinks this happens:

> We extend the place-value system by adding digits on the left, but when we read large numbers we have problems. Our brain instantly sees 3 digits as a hundred and 4 as a thousand but when working with higher numbers we have to start from the right and look carefully to see how many blocks of 3 we have to make thousands etc.

Dealing with numbers we are not accustomed to is hard, and recently Anne observed Penny, a year 2 child, struggling with the problem of writing 'forty-seven': first she wrote '40', paused and then superimposed a '7' on the '0' (see Figure 0.1). When considering the ordering and writing of numbers, one of the year 2 teachers in our group said that children frequently write '99, 100, 1001, 1002, 1003' and then read them as 'ninety-nine, one hundred, a hundred and one' etc. When Paul recounted this to the international team, it was interesting to note that similar problems were experienced in all of the project countries.

Figure 0.1 Penny's '47'

It probably does not help that we can say numbers in a variety of different ways depending on the context. Thus, for example, 2010 might be read as 'twenty ten' as a date, or 'two thousand and ten' if you are considering a number of items, or even 'two, zero, one, oh' if you are recalling part of a telephone number.

Ruth – a year 2 teacher in the research team – discussing the impact of our work, said:

> This whole project has made me very aware of zero – how important it is – and it has affected the way in which I teach but it has also reinforced some of the good things I do in terms of the visual and the practical. It makes me aware of the importance of working on zero right from day 1 and it's important to get it right otherwise it can have dire effects in year 2. I wasn't really aware of what a problem it was getting it wrong until I talked to people here. The children have got to see the practical/physical side of zero.

Possible misconceptions appearing in the later primary years

A fifteenth-century French writer described zero as 'Un chiffre donnant umbre et encombre' – a figure causing confusion and difficulty (Room, 1989: 26). In this section we will explore various mathematical misconceptions surrounding zero in the later years of primary schooling, but first we wish to share a recent observation in one of the project classrooms. Sam handed his teacher an almost complete page of problems involving the equals sign and announced 'I've finished'. The teacher, spotting $7 + \square = 5 + 2$, asked 'What about this one?', to which Sam replied 'Zero is nothing so nothing is zero'. In other words, he knew the answer was zero and chose to represent it by nothing. On reading an earlier draft of this chapter Carlo – one of the authors of Chapter 7 – was relieved for, as he explains, he had been puzzling over some work he was given to mark:

> This attitude could explain 18 pieces of work from Italian children in years 2–5. When asked to complete examples such as: $6 - \square = 8 - \square$, with two

empty squares to fill in, they only filled in one of the squares. If, however, they were assuming that an empty square represented 'zero', the task is correctly solved. Thus in the example given an answer of 2 in the second square would be correct.

One of the common problems our teachers observed was of the type $0 \times 3 = 3$. Here are some of their suggestions as to why it might occur:

> It could be that children's experience of addition and subtraction with zero means that they get use to the idea that '0' does not change anything.
>
> Pupils expect numbers to get bigger when adding or multiplying so may have problems when zero is involved in the operation since this goes against the grain. An 'empty vessel' model would be very helpful in this case. For example, you could arrange problems to include a plate with nothing on it to represent a plate of zero cakes.
>
> Or it may be a consequence of children being told 'you get more when you multiply'.
>
> Yes, but what I find surprising is that nobody in my class wrote $0 \times 3 = 30$, as I am guilty of saying 'When you multiply by ten you add a nought', and so I expected some to have overgeneralised from that.

Further insight into such issues can be gained when you slightly change the situation. For example, some children can solve $3 \times 0 = \square$ without any difficulty. Annabelle suggested that this might be because, as they had been taught that $3 \times 2 = 2 + 2 + 2$, the children converted 3×0 into a repeated addition problem, $0 + 0 + 0 = \square$, but that they were puzzled how to do so with $0 \times 3 = \square$. Ivan suggested that using the language of 'lots of' might make it easier to understand. Thus $0 \times 3 = \square$ would become 'zero lots of three equals ... '. This led the discussion on to the possibility of including zero when the children recite their tables. Thus, for example, 'zero times six equals zero, one times six equals six ...' would become the norm. As suggested earlier in the chapter, the teachers suggested that examples including zero should become far more a part of classroom life. Ruth added that

> quite a lot of maths is just picking up routines, especially if you are not very confident about the subject. It's through doing and learning the formula, and if you do something such as multiplication in different ways and you understand, then brilliant.

Some more advanced issues

Pause for thought

Many children make the following error: $5.03 + 2.1 = \boxed{7.4}$. Why do you think this might be the case?

In essence, the above is likely to be due to a misconception about place value. Asking pupils to think specifically about the value of each digit in turn is a good starting point for unravelling the problem. You might also alter the numbers to, for example, 5.03 + 5.1 or 2.03 + 2.1 and ask children to find – or even place – them on a number line to gain more of an appreciation of their values and the likely outcome of adding them together.

Our Italian colleagues described how some of their pupils seem to think that 0.07 < 0. On hearing this, several of our project teachers in the UK agreed. Donna (a year 4 teacher), however, remarked that with

> negative numbers they seem to pick up the ideas very quickly. These numbers [pointing to 1, 2, 3 on a number line] go that way [gesturing to the right] and the others go that way. We have never had any real problems. When finding the difference between −3 and +4 they never miss out the zero because they make a big deal of counting down from plus 4 to there first before going further left to minus 3.

Hearing this, Paul suggested that the 'squeeze game' might help children begin to appreciate the relationship between 0 and 0.07. In essence, you present pupils with a number line as shown in Figure 0.2. You then request someone to pick a number between –3 and +1. Suppose they pick –1, you ask them to pick a number between –1 and +1. It is likely that they will opt for 0. You then invite them to think of a number between 0 and +1, and so it goes on – possibly with some prompts from you – with the children gradually appreciating that numbers such as 0.07 might indeed be more than 0.

Figure 0.2 Using a number line as part of the 'squeeze game'

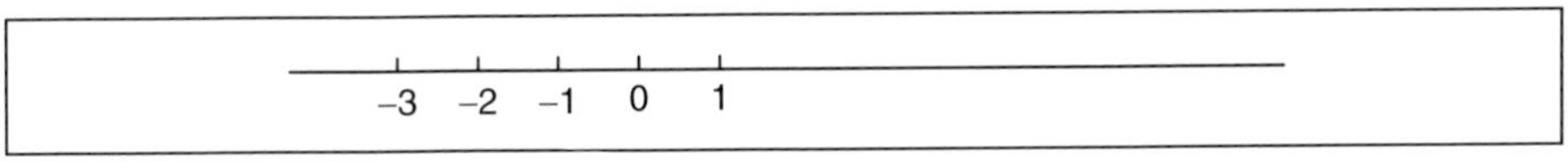

−3 −2 −1 0 1

The role of structural aids

Over the years we have seen a wide range of structural aids designed to enhance children's understanding of place value. In our view some have been misused, some helpful and some less so. In the first category was a wall chart recently seen in a year 2 classroom: it is true the numbers 0 to 20 were all there, but in such a way that it was difficult for young, insecure mathematicians to distinguish one from another.

01234567891011121314151617181920

More commonly, we have seen number fans used to good effect to display answers from 0 to 9, but confusion can arise when a child is confronted with, for example, 18 as shown below: is it meant to be eighteen or a one and an eight?

18 number fan

Several of the teachers in the study said that they found mini hundreds, tens and units flip charts very useful as they could demonstrate how a number such as 439 is built up from 400, 30 and 9.

400 and 30 and 9

439

Dienes blocks have also been found to be effective over the years, although we think they are best introduced to young children as, for example, 'pretend' items of some sort which come in packets of 10, 100 and so on. Otherwise we suspect they are a little too abstract for the lower-attaining younger children.

Some concluding thoughts

Many of the issues discussed above have been widely referred to in the literature, including a detailed discussion by Dickson *et al.* (1984) and in more recent papers by Baldazzi *et al.* (2004) and Bonotto *et al.* (2007) in Italian. We will, however, be considering some of these further in subsequent chapters, for our research has demonstrated that pupils' responses reveal much of their thinking about mathematics and that many of their misconceptions arise at an early age. In the meantime, let us leave you with some challenges.

Challenge 1

Joey, a year 2 child, was asked, 'Is the digit 0 worth more in 105 than 150? Why?' To which he replied, 'It is worth more in 105 as it is in the tens column'. How might you respond to Joey?

Challenge 2

Try the following questions posed by Carlo and Paola (authors of Chapter 7) to check on your understanding of zero (answers below):

(Continued)

(Continued)

$1 + 1 = \square$
$1 - 1 = \square$
$1 \times 1 = \square$
$0 + 0 = \square$
$0 - 0 = \square$
$0 \times 0 = \square$
$1 + 0 = \square$
$1 - 0 = \square$
$1 \times 0 = \square$
$0 \times (1 + 0) = \square$
$1 \times (1 + 0) = \square$
$(1 \times 0) + (1 + 0) = \square$

Challenge 3

Reflect on the following:

Lakoff and Núñez (2000: 75) discuss 'four grounding metaphors' to describe ways in which zero might be conceived:

> In the collection metaphor, zero is the empty collection ... In the object construction metaphor, zero is either the lack of an object ... or, as a result of an operation, the destruction of an object ... In the measuring metaphor, zero stands for the *ultimate in smallness*, the lack of any physical segment at all. In the motion metaphor, zero is the origin of motion.

Postscript

The challenge presented by Joey has initiated some lively debates: some of my colleagues say that he is right and other strongly disagree, saying that the zeros in both numbers, 105 and 150, denote an absence and therefore that they must be worth the same. On hearing of these disagreements we were tempted to omit the example, but we decided not to, thinking that it is no bad thing if people realise that mathematical ideas are not always uncontroversial! Our view is that the following explanation from a mathematician is probably closest to the truth:

> As a digit it is the same. For example, in 125 and 152 the digit '2' is the same but, when talking about quantity, there is a need to think about both the digit *and* its position. Thus the digit '2' in the context of 125 is worth more than the digit '2' in the context of 152.

The answers to Carlo and Paola's questions are as follows:

$1 + 1 = 2$
$1 - 1 = 0$
$1 \times 1 = 1$
$0 + 0 = 0$

$0 - 0 = 0$
$0 \times 0 = 0$
$1 + 0 = 1$
$1 - 0 = 1$
$1 \times 0 = 0$

Then, for the enthusiasts, remembering that you always do the calculations in brackets first,

$0 \times (1 + 0) = 0 \times 1 = 0$
$1 \times (1 + 0) = 1 \times 1 = 1$
$(1 \times 0) + (1 + 0) = 0 + 1 = 1$

You may have wondered why we did not include any division problems amongst the above. The short answer is that division by 0 is a much more complex mathematical operation than the above and simply cannot be done using natural numbers. The issue of division and zero will be discussed in more detail from a theoretical standpoint in Chapter 7 and the appendix at the end of the book, but in the meantime here is an extract from Haylock and Cockburn (2008):

> Divisions involving zero are rather more puzzling. What do you make of $0 \div 7$ and $7 \div 0$? The first can be understood using the equal-sharing structure of division: a set of zero items shared between seven people results in each person getting zero items. So $0 \div 7 = 0$. However, $7 \div 0$ makes no sense in terms of equal sharing – you cannot envisage sharing 7 items between no people! However, we can interpret it as 'how many sets of zero items make a total of seven items?' Well, you can go on accumulating sets of zero items as long as you like and you will never reach a total of seven. For example, we have put two thousand sets of zero elephants inside these brackets: [ ] – and we're still nowhere near to getting seven elephants! The conclusion of this line of reasoning is that you just cannot do $7 \div 0$. It is a calculation without an answer. Mathematicians say that division by zero is not allowed. So does any calculator: try it and see. (You may have heard somewhere that $7 \div 0$ is equal to infinity – there is a branch of number theory that uses that kind of language, but please do not think that there is a real number called 'infinity'.) So, in summary, for any number a, the following generalizations can be made for multiplications and divisions involving zero:
>
> $a \times 0 = 0$
> $0 \times a = 0$
> $0 \div a = 0$ (provided a is not zero)
> $a \div 0$ cannot be done

Further reading

When we looked through the books we generally recommend to students and teachers we found very few references to zero, and then only a page at most. We then did a Google Scholar search, only to find numerous, but highly advanced, references. Having said that, looking in the library, we found Robert Kaplan's (1999: 1) book, which begins:

> If you look at zero you see nothing; but look through it and you will see the world.

The book is entitled *The Nothing That Is: A Natural History of Zero*. It has been described as 'A true delight' (Roger Penrose) and 'Entertaining, informative, brilliantly done' (Martin Gardner), and it is certainly worth reading if you wish to learn more about the origins of zero.

I came across another book, *The Invention of Zero* by Chris Greenhalgh (2007: 9), which we thought might be relevant, but it proved to be a poetry book. Having said that, the first and last verses of a poem by the same title struck a cord:

> Nature adores a vacuum –
> a few bits of light
> snagged on nothingness
> ...
> the moon –
> white beyond bleaching,
> the end of abstraction;
>
> a perfect blank.

References

Baldazzi, L., Cottino, L., Dal Corso, E. *et al.* (2004) Le competenze dei bambini di prima elementare: un approccio all'aritmetica. *La Matematica e la sua Didattica*, 18(1), 47–95.

Bonotto, C., Baccarin, R., Basso, M. and Feltresi, M. (2007) Sulle strategic di conta e procedure di calcolo mentale in bambini di scoula primaria. *L'Insegnamento della Matematica e delle Science Integrate*, 30A(5), 519–548.

Dickson, L., Brown, M. and Gibson, O. (1984) *Children Learning Mathematics*. London: Cassell.

Flegg, G. (ed.) (1989) *Numbers Through the Ages*. Basingstoke and London: Macmillan Education in association with the Open University.

Greenhalgh, C. (2007) *The Invention of Zero*. Tarset: Bloodaxe.

Haylock, D. and Cockburn, A.D. (2008) *Understanding Mathematics for Young Children*. London: Sage.

Ifrah, G. (1998) *The Universal History of Numbers*. Translated from French by D. Bellos, E.F. Harding, S. Wood and I. Monk. London: Harvill Press.

Kaplan, R. (1999) *The Nothing That Is: A Natural History of Zero*. London: Allen Lane.

Lakoff, G. and Núñez, R.E. (2000) *Where Mathematics Comes From*. New York: Basic Books.

Room, A. (1989) *The Guinness Book of Numbers*. London: Guinness Books.

In addition, the teachers in conversation, mentioned the following:

Department for Education and Employment (1999) *The National Numeracy Strategy, Framework for Teaching Mathematics from Reception to Year 6*. Sudbury: Department for Education and Employment.

Department for Education and Skills (2006) *Primary Framework for Literacy and Mathematics*. London: Department for Education and Skills.

Qualifications and Curriculum Authority (1999) *Early Learning Goals*. London: Qualifications and Curriculum Authority.

1

Equality: getting the right balance

Paul Parslow-Williams and Anne D. Cockburn

> **Pause for thought**
>
> What do you think of when you see this sign '='? If you had to discuss it with a class of children, what would you say? Would you be tempted to describe it as a symbol for 'makes'? How about 'the same as'? Do you always write number sentences with this sign at the end?

When we had a look at what people have been writing about the equals sign (see, for example, Behr *et al.*, 1980; Falkner *et al.*, 1999; Freiman and Lee, 2004) we discovered it is well documented that children frequently find it difficult to appreciate that:

- '=' signifies 'the same as', but not necessarily 'identical to'.
- The equals sign is not a request to *do* something: '+' invites you to add items, '–' asks you to subtract but '=' simply states the situation rather than demanding any action.

As Jones (2006: 6) puts it: 'An arithmetic expression is like a film set on which the numbers are actors, the operators are the script and the equals sign the director who shouts "Action!"' If you are an experienced teacher, you were probably not surprised to read this, but we hope that, like us, you feel that simply knowing the misconceptions children hold is not enough and are intrigued to delve further into why children experience these particular difficulties. To this end, in this short chapter we will investigate some of the most common misconceptions we observed around the equals sign. They will illustrate the above but, more significantly for our purposes, they provide insight into how children try to make sense of the world of mathematics and how we might better help them to do so. Much of the material discussed arises from children's responses to the equality problems presented in Figures 1.2 and 1.3. Before we turn to these, however, let us take a few steps back in time.

A brief history of the equality symbol

Whilst we may take the equality symbol '=' for granted as part of our everyday mathematical vocabulary, it was not until the sixteenth century (1557) that this form, albeit in a rather elongated version, was first seen in print in Robert Recorde's *Whetstone of Witte* (a whetstone is a device for sharpening tools and, in the title of Recorde's book, it is assumed that the 'witte' being honed is one's mathematic understanding). Prior to this, equality had been symbolised in a variety of ways, including the 't' used in the third-century manuscripts of Diophantus, '*ae*' (an abbreviation for the Latin word *aequalis*) and a pair of vertical lines '||'(Cajori, 1928; Saenaz-Ludlow and Walgamuth, 1998).

Recorde (1557) rather elegantly justifies its use, explaining (see Figure 1.1):

Figure 1.1 Extract from Robert Recorde's *The Whetstone of Witte*, 1557 (Cajori, 1928: 165)

The Arte

as their workes doe extende) to distincte it onely into twoo partes. Whereof the firste is, *when one nomber is equalle unto one other*. And the seconde is, *when one nomber is compared as equalle unto.2.other nombers.*

Alwaies willyng you to remẽber, that you reduce your nombers, to their leaste denominations, and smalleste formes, before you procede any farther.

And again, if your *equation* be soche, that the greateste denomination *Cosike*, be ioined to any parte of a compounde nomber, you shall tourne it so, that the nomber of the greateste signe alone, maie stande as equalle to the reste.

And this is all that neadeth to be taughte, concernyng this woorke.

Howbeit, for easie alteratiõ of *equations*. I will propounde a fewe exãples, bicause the extraction of their rootes, maie the more aptly bee wroughte. And to avoide the tediouse repetition of these woordes: is equalle to: I will sette as I doe often in woorke use, a paire of paralleles, or Gemowe lines of one lengthe, thus: ═══, bicause noe.2. thynges, can be moare equalle. And now marke these nombers.

1. 14.ze. + .15.ϑ = 71.ϑ.
2. 20.ze. — .18.ϑ = .102.ϑ.
3. 26.ʒ. + 10ze = 9.ʒ — 10ze + 213.ϑ.
4. 19.ze + 192.ϑ = 10ʒ + 108ϑ — 19ze
5. 18.ze + 24.ϑ. = 8.ʒ. + 2.ze.
6. 34ʒ — 12ze = 40ze + 480ϑ — 9.ʒ

1. In the firste there appeareth. 2. nombers, that is 14.ze.

> To auoide the tediouse repetition of these woordes : is equalle to : I will sette as I doe often in woorke vse, a paire of paralleles, or Gemowe* lines of one lenghte, thus: ═══, bicause noe.2. thynges, can be moare equalle.

Despite his significant contributions to mathematics, which also include the introduction of algebra to the British Isles and authorship of a series of mathematical texts in the English language making geometry and astronomy accessible to a wider audience, Recorde is relatively unknown. If you would like to find out more about this fascinating Welshman, including his rise and fall in Tudor politics and how he came to a rather unfortunate end as a pauper in jail, there are a number of short biographies to be found on the Internet (see, for example, his entry in the online Encyclopædia Britannica). If you wish to focus on his mathematical accomplishments in a wider context, you will enjoy reading Cajori (1928).

However, before we move on, it is worth noting that although Recorde is usually credited with being the first to use this particular symbol, there is some evidence in the form of a manuscript from the University of Bologna that suggests that it may also have been developed elsewhere (Marchini, personal communication, 2007).

Challenge

Ask your class – regardless of their age – to complete Figures 1.2 and 1.3.

Figure 1.2 A series of equality problems based on addition

Can you complete these number sentences?

a) 7 + 2 = □

b) 5 + □ = 8

c) □ + 4 = 9

d) □ = 3 + 4

e) 5 = □ + 1

f) 8 = 5 + □

g) 5 + 4 = □ + 8

h) 6 + 2 = 3 + □

i) 1 + □ = 6 + 2

j) □ + 3 = 7 + 2

k) 5 + □ = □ + 7

l) 9 = □

m) 5 + 4 = □ + 6 = □

n) 4 + 3 = 2 + □ = □ + 1 = □

*'Gemowe' means 'twin'.

Figure 1.3 A series of equality problems based on subtraction

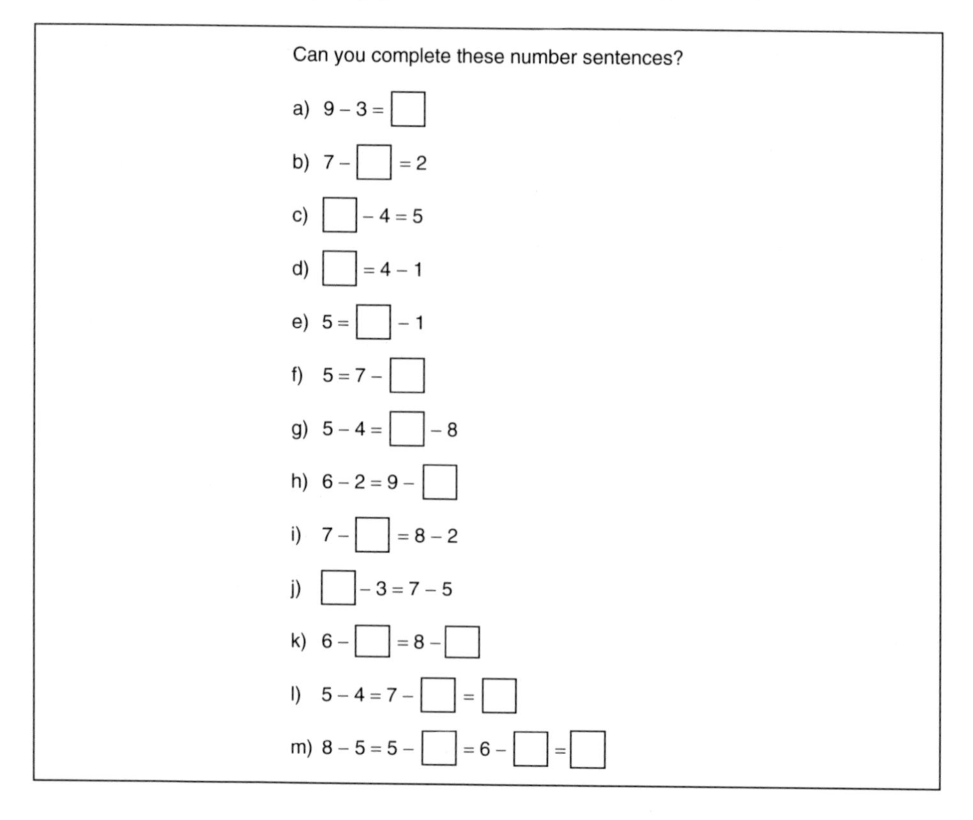

Ivan, a year 1 teacher, presented Figures 1.2 and 1.3 as 'problems of the week' and invited the children to 'have a go' and to stop when they wished. He noted that children were more likely to attempt the problems involving the '+' sign as opposed to those including the '–' sign. Furthermore, their success rate was higher for addition than subtraction. From this he concluded that the children were 'obviously more confident in their own adding ability than subtracting'. Ivan may be correct that this was his pupils' perception but, interestingly, to solve a tricky problem such as 5 = □ + 1 it is highly likely that the children would have used one of the following strategies:

- a 'trial and improvement' approach that involved inserting different numbers in the place of the unknown until balance was achieved, possibly thinking along the lines of 'What do I need to add to 1 to make 5?' or
- subtraction.

Similarly, almost all of the class successfully solved 5 + □ = 8, for which they almost certainly used 'counting on' to 8. The initial point we wish to make here is that, although the use of a particular sign (in this case '+') may have given the children more confidence as Ivan suggested, it did not necessarily mean that they were performing the simple operation one instinctively associates with the '+' sign.

The second point from the above is to question the notion that Ivan's class saw '=' as a command for action which, coupled with '+' or '–' in these cases, was to add or subtract. Certainly a couple of children did respond $5 = \boxed{6} + 1$ but none of them answered $5 + \boxed{13} = 8$. Further support for this idea comes from the observation that, although over half of them successfully inserted '9' in $\square - 4 = 5$, only one child responded '1'. Comparing the above with the responses of Italian children of similar age, it is interesting to note that over 90 per cent answered $5 + \square = 8$ correctly, implying that also for them '=' together with '+' does not necessarily involve the need in children's minds to add 5 and 8.

Looking at the responses to the more complex problems in Figures 1.2 and 1.3 provides further insight into children's perceptions of '='. Take $5 + 4 = \square + 8$ for example. Half of the year 1 children correctly responded '1', but a quarter of them inserted '9' (a comparable proportion of Italian children responded similarly). If we also consider $5 - 4 = \square - 8$, we find that only one of the 22 children in Ivan's class was correct while almost half of them gave the answer as '1'. Similarly, many Italian pupils in years 2, 3 and 4 also responded with '1'. Ivan explained one of the most likely reasons for these responses in terms of Adam's answer:

> He has put 1 when it should be 9. I think he has seen the bit he knows, i.e. $5 - 4 = 1$, because that is the bit he could do and there is this perception that the answer always comes after the equals sign (he ignored the –8). When I spoke to him about it he didn't understand it.

It may be that, when uncertain, children will tackle what they are familiar with in an endeavour to do as much of the work as they can for you. Theo (aged 6), for example, produced the following: $7 - \boxed{13} = 8 - 2$. In effect he took 2 from 8 and arrived at 6 and then worked out what needed to go into the box to produce a balance of 6 on the left-hand side of the equation. The idea of balance is taken up again later in the chapter.

You may, like some of the teachers in the project, feel that $9 = ?$ is a rather unconventional question to present to children, yet many of the responses we collected from pupils of all ages were illuminating. Thus, for example, Renato, a year 2 child in Italy, believing that the equals sign should always be coupled with an operator such as '+' or '×', wrote $9 = 0$ when confronted with $9 = \square$, explaining 'I wrote zero because there is nothing to do with the 9'. A year 6 child produced an equally unexpected but valid response, writing $9 = 27$ and then divided the response box into two to indicate '2' and '7' as separate entities. His teacher suggested that:

> Many (even Level 5) struggled with $9 = \square$ or $42 = \square$. Pupils are so accustomed to having to do something.

None of the children in Ivan's class showed any of their workings when attempting to solve the problems. However, when Ruth asked seven of her

Figure 1.4 Examples of pupils' recording strategies for solving equality problems

$6 - 2 = 9 - ?$

////XX ///XXXXX

$? - 3 = 7 - 5$

/XXX //XXXXX

$5 = ? - 1$

////X

pupils to complete the subtraction problems they all included tally marks to indicate their thinking. As Ruth explains:

> They've been encouraged from day 1 to show their working. They've been drilled into it. They are so used to using structural apparatus that when there are no cubes available they do sticks and dots.

Their strategy was typically to set out sticks underneath on either side of the equals sign and score out the sticks to be subtracted (see Figure 1.4). Ruth's children – at the age of 6 – were a year older than Ivan's, but it is interesting to compare their responses in the light of their differing approaches. When confronted with the problems in Figure 1.4, the recording strategy initially proved productive for Ruth's class, who were able to answer the first two questions correctly, whereas few of Ivan's children found the correct answers. In the first, the left-hand side provided a pattern, so the pupil had only to make the nine strokes and cross out sticks one at a time until they were left with a similar pattern of four uncrossed sticks. The middle example arguably demanded a much higher order of thinking than the other two; whilst the right-hand side gave the pattern, it was not immediately clear to the pupils how many sticks to draw. To reason that three crossed sticks of a yet unknown number must be placed and then to put two uncrossed sticks on the end (to match those on the other side) was quite a difficult process. But note that none of those who did workings succeeded with the third question. All the year 2 children drew five tallies (/////) and crossed out one (////X). It was not uncommon in the project to find that children experienced problems when attempting to apply naïve strategies to situations that differed from the context in which they were introduced. Even a small change, such as the number of terms in the question, was often sufficient to confuse many of the children. For a theoretical framework and practical advice on supporting pupils with developing their mathematical strategies, please refer to Chapter 6.

Challenge

Consider how you might complete the following: 6 ? 1 ? 5 (each ? represents a mathematical symbol). Is there more than one possible solution?

As we read through the children's responses we frequently gained the impression that they were trying to make sense of what they had been asked to do. The most striking examples of children seemingly applying their own logic were revealed by problems such as □ = 3 + 4 and □ = 4 – 1, to which many children responded with the answers of 1 and 5, respectively. It is possible that these children, in effect, did the opposite operation to the one required and made us question whether, on seeing a problem written 'reversed', they assumed that their task was also reversed – to add rather than subtract and vice versa.

Alternatively, might the children have been ignoring the = and +/– signs in order to force a left to right operation? To take □ = 3 + 4, you could also arrive at 1 by thinking of the calculation as 1 '+' 3 '=' 4. This also holds true for the second example, since 5 will also be the result if □ = 4 – 1 is considered as □ – 4 = 1. In both of these cases, perhaps the position of the symbol is more important than what it represents. Another possibility is that the position of the symbol is irrelevant in the minds of some of the children but its function is crucial.

Challenges

What errors do your pupils make when doing problems involving the equals sign? Do you notice any pattern to their mistakes? What misconceptions might these reveal?

Contributory factors

Not to put too fine a point on it, the project teachers were amazed by their pupils' responses to our equality problems. As they explained, previously, they had not perceived the seemingly innocent '=' sign as an issue:

> We never discuss explicitly what this sign means and I feel that this is something that as a school we need to look at now. (Linda, year 2 teacher)
>
> We didn't talk about the equals sign and what it meant at teacher training. (Kath, year 6 teacher)
>
> We use the equals sign all the time but never talk about what it means. We don't make the connections when we look at it. (Laura, year 3 teacher)
>
> We teach the inequality signs but there is little emphasis on the equals sign. The children would probably find ? > 3 + 4 easier than ? = 3 + 4 (Donna, year 4 teacher)

Previously lacking an appreciation of the equals sign, Kath (year 6 teacher) confessed that she was probably responsible for some of her pupils' errors such as the example shown below:

Problem: 48 – □ = 47 – □ = 46 – □ = □

Pupil response: 48 – [1] = 47 – [1] = 46 – [1] = [45]

It appears that the pupil interpreted the problem as a chain of commands leading to a final answer rather than one long number sentence with three equality symbols. Kath went on to explain that, in the past, she had regularly used the equals sign when it is not mathematically appropriate. Thus, for example, in order to demonstrate partitioning strategies to pupils, she might have written

$$45 + 22 = 40 + 20 = 60 + 5 = 65 + 2 = 67$$

The project made her

> think about how I write on the board. I should use arrows more. Today has been useful from my point of view as a teacher. I do lots of things subconsciously that are causing misconceptions. Now I'm going down to a younger year group I have a better understanding of what I need to be really careful about.

Seeing results from both his own and other classes, Ivan also concluded that some of his teaching strategies might have contributed to children developing misconceptions which, although not apparent in the short term, created difficulties at a later stage in their schooling. He reflected:

> You focus on your own group and forget what they are going on to. I've done year 3,2,1 and R and so you get an idea of what the children are moving towards ... [but] some of the children have misconceptions and still have them 5 years on.

(See Chapter 6 for actual examples of the same misconceptions in year 2 and year 10.) Ivan further explained:

> You tend to cover things very quickly because of the expected coverage when the children are not getting the bedrock. I have done 'if you add 2 things it gets bigger and if you subtract it gets smaller'. You don't tend to think what you do has bearing on what others do.

This mirrors findings in another research project and is reported in Cockburn (2007).

Challenge

What might you do which could contribute to your pupils' misunderstanding of equality?

Some ways forward

Clearly children have to develop a real appreciation of the meaning of mathematical equality which, traditionally, is represented by '='. This is not just to complete simple addition, subtraction, multiplication and division problems and not just to tackle the vaguely familiar subject of algebra which, to an early years teacher, may seem so far into a child's future as to be almost irrelevant. Indeed, reflecting on the challenge below, you will note that we expect children to have a fairly developed sense of both the composition of numbers and the relationships between them by the time they are engaging in what traditionally was termed 'mental arithmetic' in year 1.

Challenge

Reflect on what you have read in this chapter and the implications it might have for teaching the following relationships:

$7 + 5 = (2 + 5) + 5 = 2 + (5 + 5) = 2 + 10 = 12$
$24 = 20 + 4\ldots$

Rather than viewing young children as 'empty vessels' who need to acquire a mass of skills before they can advance mathematically, we think it would be helpful to adopt a different perspective at this point and consider what they can do, rather than what they cannot.

For years, for example, it has been recognised that children as young as 3 years old can successfully complete a range of numerical problems – including division – if presented in an appropriate manner using real objects within a familiar everyday context (see, for example, Desforges and Desforges, 1980; Hughes, 1986; Gelman and Gallistel, 1978). Your own experience may also tell you that, when totting up the numbers on the register, Key Stage 1 children can solve problems which, when written formally, might be mathematically represented by $24 - \square = 21$ or even $15 + \square = 21$, if you were considering how many children had opted for packed lunches and the number of hot dinners needed to be calculated. We will return to this later, but first let us consider another typical aspect of young children's mathematical knowledge.

In addition to asking pupils to complete formal equality problems as presented in Figures 1.2 and 1.3, we asked them to undertake a series of tasks to explore their understanding of equality in a visual sense, as illustrated in Figure 1.5.

Sandra asked five of her reception class to complete the exercise. Three of the 5-year-olds successfully completed (A) independently by drawing two blocks on the right-hand pan, explaining:

> That's got 3 and that's got 3. (Kareen)
>
> Now it will balance. (Lorna)
>
> I drew 2 blocks to make it the same. (Ron)

Neil and Miranda also achieved success but adopted unexpected strategies. Neil began by drawing three cubes on the left-hand pan and then drew a further 5 on the right hand side and explained 'Six and six'. Entirely separately, Miranda drew 5 blocks on the left hand side and 7 on the left and announced, 'Eight on each side'.

Figure 1.5 Equality problems represented in a visual form

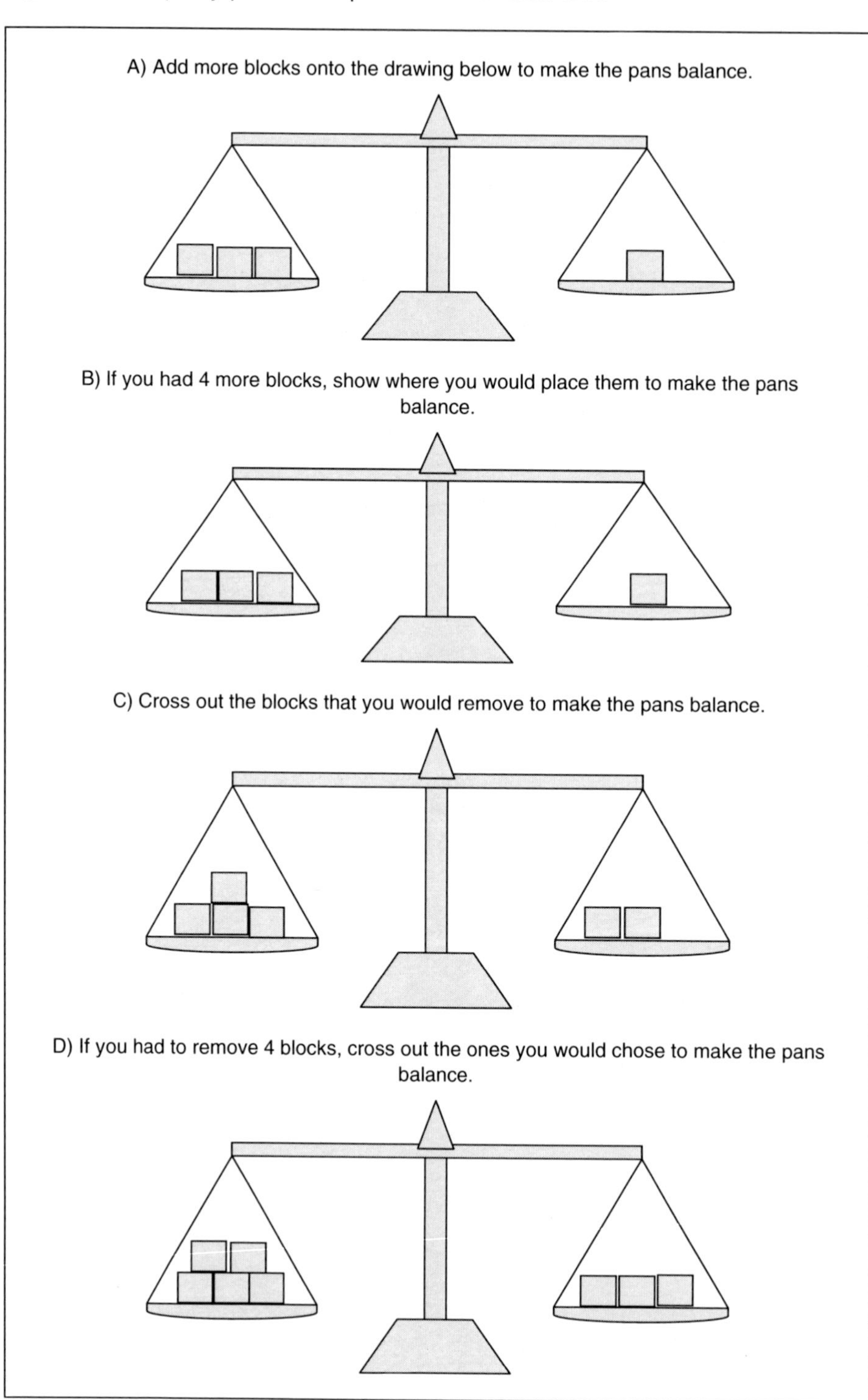

The children's responses to the other tasks were fairly standard, with their comments suggesting a good understanding of equality, such as 'Now there is 4 on each side' (Kareen) and 'I've made 2 and 2' (Ron). It is interesting to note though that, when completing (C), both Neil and Lorna – again independently – crossed out all of the cubes in sight. Viewing the result, Neil commented, 'Nothing on each side', and Lorna remarked 'Now there's none on each side so they [the pans] will be the same height'.

The children's teacher was not surprised, as she said that she has noticed over the years that reception children generally find conceptions surrounding weight straightforward and easy. She continued:

> They readily understood the balance so let's start to push their thinking out ... It's building on previous experience which has been fairly well digested. It's giving them another hook into the learning: a visual hook. It's giving them another way in. Verbal language doesn't necessarily unlock doors for them.

A year 2 teacher from another school, Linda, described how she had in the past used:

> balances specifically to demonstrate equality. This is a very powerful method for introducing this concept. Three cubes on one side, one on the other. Pupils can see (since one side is lower) that these are not equal. We used to do lots of work on equality and talked about balancing using apparatus where appropriate, but we haven't done a lot of this since the introduction of the Strategy. Looking through these examples has made me think about my teaching of equality. When looking at the tasks I thought 'I don't do this sort of thing with my children'.

She went on to say, however, that her

> pupils found it difficult when balances were on paper. When the children were asked to add four cubes (question (B), they wanted to put the four cubes on one side. They didn't understand that they could split the set. The children expected the solutions to be simpler and wanted to solve them quickly and at a superficial level.

Kath – a year 6 teacher – also referred back to her previous practice when she saw the results of the task:

> I wish we still had the old equaliser balances since when we went back to showing them the balances, they found it much easier to understand ... [at the moment] they understand equals as makes rather than a balance.

So, to return to young children's facility to solve realistic problems: have we been seduced into thinking that the transition to formally recording such situations is easier than it actually is? We know from the work of others that they can record their work informally using symbols and pictures (e.g. Hughes, 1986). We have observed that children as young as 5 understand the practical concept of balance and hence equality. Might the two approaches be combined so that pupils were encouraged to undertake their calculations and illustrate their thinking perhaps using, in the first instance, real objects such as marbles or miniature bananas and classroom balances?

Working with a top-pan or suspension balance offers children a concrete experience of equality and an opportunity to engage physically with number sentences. Thus, for example, the following conversation might take place:

Teacher: If I have 2 bananas and you have 3, how many would you need to eat so that you have the same number of bananas as I have? How could we show it on the balance?

Sam: You could put your 2 bananas on your side and I could put my 3 on mine. That makes my side lower but if I take one off my side – I could pretend to eat it! – then it balances.

To take things a step further, drawstring bags could be used to represent unknown sets. For example, to demonstrate the problem $3 + \square = 8$, a bag containing five hidden marbles would be placed in the left-hand side of the balance with three visible ones, and eight marbles would be placed in the other (to balance the scales, an empty bag would also need to be placed in the right-hand pan); see Figure 1.6. Although the potential exists to use this model to develop algebraic reasoning with upper primary pupils, at a basic level it provides a visual representation of the 'missing number' type questions (such as those in Figures 1.2 and 1.3) that can be explored by young children.

Figure 1.6 Visual representation of a 'missing number' equality problem using a suspension balance

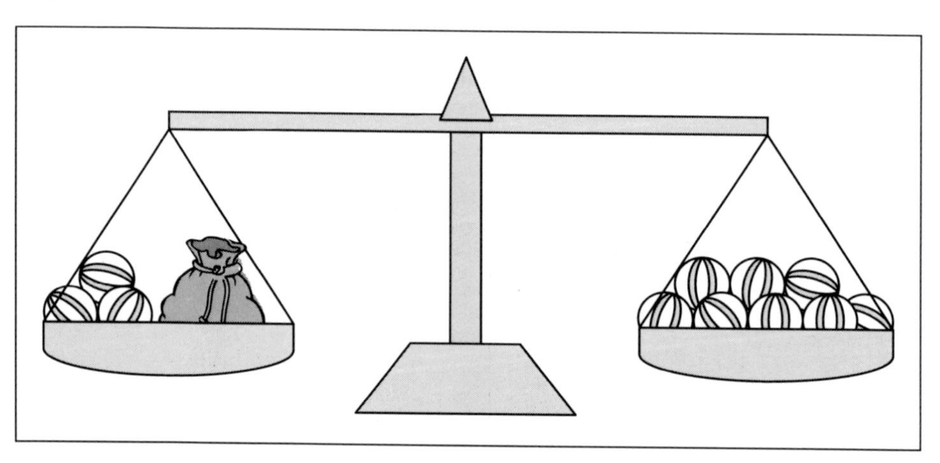

To begin the move towards symbolic representation, this idea could then be extended to cards on which the objects are represented pictorially (see Figure 1.7). Following on from this, pupils would be encouraged to make their own marks to record, eventually leading to the written form.

It is of fundamental importance that we consider the difficulties pupils experience with equality when selecting visual images and apparatus to illustrate operations – subtraction in particular. Instead of showing 'before and after' models with the subtracted objects removed completely in the 'after'

Figure 1.7 Picture cards used to represent objects in the equality relationship 2 = 3 –1

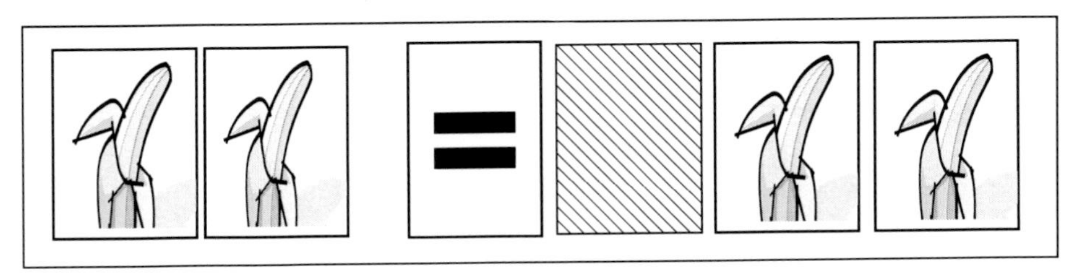

image, it is helpful if we leave a symbolic representation of them behind so that children can see the whole number sentence and remind them of what operation has taken place in order to achieve balance. Even modest changes to practice such as crossing out instead of erasing the 'eaten' sweets from the board, flipping over rather than removing picture cards (as in the example in Figure 1.7) or drawing empty lily pads to represent missing frogs (see in Figure 1.8) could potentially have profound influences on pupil understanding.

Figure 1.8 An example of a pictorial model that could be used as a basis for discussion of number sentence structures

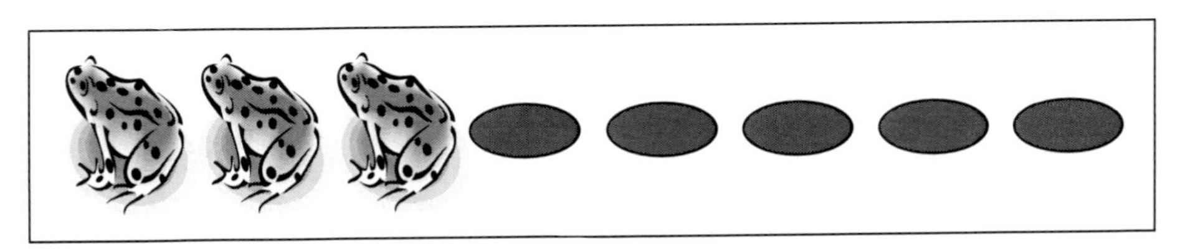

In Figure 1.8, a complete set of 8 can be seen (8 lily pads) as well as the partitioning into subsets of 3 (occupied) and 5 (unoccupied). With careful use of language such as 'How many more frogs would be needed so that all of the lily pads are covered?' or 'If there were 8 frogs to start with, how many have hopped away to leave 3?', images similar to the one in Figure 1.8 could be used a basis for discussing problems such as 3 + □ = 8 or 8 – □ = 3 and, in doing so, help pupils to develop their familiarity with a range of number sentence structures.

Concluding remarks

The notion of equality is a central – but sorely neglected – concept in mathematics education. There is considerable scope to introduce children – albeit often unwittingly – to strategies which have the potential to create misconceptions in their thinking and thus cause difficulties in their future mathematical development. What is more, studies on older pupils (e.g. Knuth *et al.*, 2006; Kieran, 1981) have identified a strong relationship between understanding of the equals sign and the ability to solve algebraic equations. Thus, perhaps primary teacher education should not confine its mathematics curriculum to a narrow age range: Kath, for example, explained that equality 'wasn't taught to me as I was trained for upper primary'. Ian also suggests that there is a 'case for at least getting more talking between year groups in school'.

Many of us are tempted to use a particular tried and tested model or phrase which we know from our own experience works very effectively for children of a particular age. Indeed, using analogies and endeavouring to simplify situations is a natural, and entirely reasonable, part of the teaching process. However, teachers need to have a greater understanding of the broader mathematical context in which they do this. Overemphasis of a particular model can make it more difficult for children to solve problems when they are presented in unfamiliar contexts. Sandra (year 2 teacher) stressed the importance of varied and accurate 'teacher talk', commenting that:

> Children have better understanding if teachers use a range of vocabulary associated with the equality symbol such as the 'same as' as well as 'total'.

This is supported by observations made by other contributing authors in this book who noted that children had a better understanding of equality after teachers had encouraged them to articulate number sentences just as they would read sentences in literacy. Ruth (also a year 2 teacher) highlighted the need for teachers to vary the position of the equality symbol and avoid always setting out number sentences in the same 'left to right' format $a + b = c$, adding

> the further up the school they go the more they will have to get use to seeing it in different ways.

Such issues are discussed in more detail in Chapter 5.

Summary of key ideas

- Children see '=' as an instruction to complete an operation. Emphasise the equivalence aspect of the '=' symbol by using phrases such as 'is the same as' or 'gives the same result as' in preference to 'makes' or 'leaves'.
- Use concrete apparatus such as balances and visual images to represent a variety of number sentence structures with the 'unknown' on both the left- and right-hand sides of the equals sign.
- Take care with how you use the '=' sign when demonstrating complex problems with multiple steps. Use arrows if it is necessary to link the successive stages together.
- Talk to colleagues teaching older and younger primary classes. What mathematical misconceptions are commonly held? Try to discover their origins and work together to develop strategies to prevent their occurrence and perpetuation.
- Children can be very innovative when presented with unconventional problems, and their responses can reveal much about both your teaching and their mathematical understanding.

Further reading 

Haylock, D. and Cockburn, A. (2008) *Understanding Mathematics in the Lower Primary Years*. London: Paul Chapman Publishing.

Chapter 1 of Haylock and Cockburn explores the various ways in which children interpret the equals sign and pays particular attention to the notions of transformation and equivalence as well as offering practical activities to use in the classroom.

Anghileri, J. (2000) *Teaching Number Sense.* London: Continuum.

The third chapter of this very accessible text provides interesting insights into how children develop their understanding of mathematical symbols.

Cajori, F. (1928) *A History of Mathematical Notation: Notations in Elementary Mathematics* (Vol. 1). Chicago: Open Court.

Although 80 years have passed since it was first published, Florian Cajori's two-volume work continues to be reprinted and has been described by a reviewer on Amazon.com as

> unsurpassed ... this history of mathematical notation stretching back to the Babylonians and Egyptians is one of the most comprehensive written. ... Florian Cajori shows the origin, evolution, and dissemination of each symbol and the competition it faced in its rise to popularity or fall into obscurity.

This is certainly not an 'entry level' text by any means, but I would recommend this book to mathematically inclined readers who also have an interest in history.

References

Behr, M., Erlwanger, S. and Nichols, E. (1980) How children view the equal sign. *Mathematics Teaching,* 92, 13–15.

Cajori, F. (1928) *A History of Mathematical Notation: Notations in Elementary Mathematics* (Vol. 1). Chicago: Open Court.

Cockburn, A.D. (2007) Understanding subtraction through enhanced communication. In A.D. Cockburn (ed.) *Mathematical Understanding 5–11.* London: Paul Chapman Publishing.

Desforges, A. and Desforges, C. (1980) Number-based strategies of sharing in young children. *Educational Studies,* 6, 97–109.

Falkner, K.P., Levi, L. and Carpenter, T.P. (1999) Children's understanding of equality: A foundation for algebra. *Teaching Children Mathematics*, 6, 232–236.

Freiman, V. and Lee, L. (2004) Tracking primary students' understanding of the equality sign. In M. Hoines and A. Fuglestad (eds), *Proceedings of the 28th Conference of the International Group for the Psychology of Mathematics Education* (Vol. 2, pp. 415–422). Bergen: Bergen University College.

Gelman, R. and Gallistel, C.R. (1978) *The Child's Understanding of Number.* Cambridge, MA: Harvard University Press.

Hughes, M. (1986) *Children and Number.* Oxford: Blackwell.

Jones, I. (2006) The equals sign and me. *Mathematics Teaching,* 194, 6–8.

Kieran, C. (1981) Concepts associated with the equality symbol. *Educational Studies in Mathematics,* 12, 159–181.

Knuth, E.J., Stephens, A.D., McNeil, N.M. and Alibali, M.W. (2006) Does understanding the equal sign matter? Evidence from solving equations. *Journal for Research in Mathematics Education,* 37, 297–312.

Saenz-Ludlow, A. and Walgamuth, C. (1998) Third graders' interpretations of equality and the equal symbol. *Educational Studies in Mathematics,* 35, 153–187.

Beginning to unravel misconceptions

Sara Hershkovitz, Dina Tirosh and Pessia Tsamir

A primary mathematics lesson in Israel

Previous chapters have discussed some of the underlying mathematical misconceptions commonly encountered in primary schools. In this chapter we wish to explore how, in the midst of a busy classroom environment, you can begin to unravel some of the misconceptions often associated with the tricky topic of place value. We will also discuss how our thinking evolved on such matters as it provides further insight into the complexities of the issues together with, we hope, some practical suggestions from the teachers involved in our project. But first a challenge:

Challenge

Look at the following symbols. Which number systems do they represent? What is the value of the number that they all show?

a. CCCLXV
b. שסה
c. 𓍢𓍢𓍢𓎆𓎆𓎆𓎆𓎆𓎆𓏺𓏺𓏺𓏺𓏺

All of the above symbols represent the number 365: CCCLXV – is in the Roman system, שסה in the Hebrew system and 𓍢𓍢𓍢𓎆𓎆𓎆𓎆𓎆𓎆𓏺𓏺𓏺𓏺𓏺 in the Egyptian system.

One of the important differences between the above and 365, a Hindu-Arabic number, is the use of the decimal system. This system – used by most of the world – utilises a place-value principle which can be extremely helpful when adding, subtracting, multiplying or dividing. When we write the number 365, for example, we use 3 digits – 3, 6, 5 – as shorthand for $3 \times 100 + 6 \times 10 + 5 \times 1$. Similarly, with the number 5007 we mean $5 \times 1000 + 0 \times 100 + 0 \times 10 + 7 \times 1$. Such an efficient use of only ten digits, in this case 0–9, is one of the great strengths of the Hindu-Arabic system and it makes numbers easier to manipulate and mathematical calculations considerably quicker to complete successfully. We are all familiar with using the system and between us, over the years, we have probably done millions and millions of calculations.

Simple though it may appear, it is relatively easy for children to pick up the wrong end of the stick and become confused over the meaning of digits in different positions. The situation becomes further complicated if they are asked to manipulate numbers in the pursuit of an answer to a calculation they have been set. Some pupils have a shaky understanding of place value, some find calculations problematic, and some find both a major difficulty. Our challenge as teachers is to find out exactly what each individual finds tricky and act accordingly. As we know only too well, however, children have an uncanny way of happily getting on with their work and sometimes it can be incredibly hard in a lively and busy lesson both to spot problems and to unravel their origins. At the outset of our project we devised some tasks to help address some of the problems and, as you will observe, during the course of our work teachers further expanded and refined them. At a glance the remainder of the chapter might look rather repetitive: it is not! It is true that what we discuss revolves around very similar tasks but, as you will discover, the detail reveals a huge amount about children's mathematical knowledge and understanding.

The original task

The original task we devised is set out in Figure 2.1, and we asked groups of primary teachers to try it out with their pupils in each of the four project countries (Italy, the Czech Republic, Israel and the UK).

Figure 2.1 Arranging four digits to create a range of calculations

Given the digits 2, 5, 6, 8, place them so as to obtain the largest result.
You may use each digit only once in each calculation.

____ ____ + ____ ____ = ______
____ ____ − ____ ____ = ______
____ ____ × ____ ____ = ______
____ ____ ÷ ____ ____ = ______

Before reading about our findings, you may wish to try the task and reflect on which aspects you found the hardest. Do you think the same would be true for your class? What kind of errors do you think might arise? Why? If it is possible – and we entirely appreciate that it might not be – it would also be useful if you are able to present these problems to an upper primary class. It would be interesting to make comparisons between the approaches taken by your class and those by children involved in the project.

The teachers in our project were very surprised by the children's responses: they had predicted that they would be more successful with addition and multiplication rather than subtraction and division. Contrary to expectations most of children (aged between 10 and 11 years) solved the subtraction and the division problems correctly, finding the most difficult task to be the one involving multiplication. The addition task proved to be more difficult than the subtraction, but less difficult than the multiplication task.

Pause for thought

Having solved the problems, can you explain the above findings? Why do you think the project children had greater success with subtraction and division than with addition and multiplication? Did you find the same with your class? How might any similarities or differences be explained?

The addition task

To understand these findings better, let us begin by taking a closer look at the addition task as given in Figure 2.2. Below we provide a fairly detailed account

Figure 2.2 Arranging four digits to create a calculation involving addition

Given the digits 2, 5, 6, 8, place them so as to obtain the largest result.
You may use each digit only once.

____ ____ + ____ ____ = ______

of project children's responses, in part so that, if applicable, you can compare them with data from your own class, and in part so that you can begin to appreciate the value of different components of the task.

A few pupils gave more than one correct answer such as:

62 + 85 = 147
85 + 62 = 147
82 + 65 = 147
65 + 82 = 147

The majority who answered correctly, however, gave only one answer and usually explained their thinking thus:

> The largest number I put in the tens place, and the other number I put in the units place. (Sally)

> I looked for the two largest numbers and I put them at the beginning of the 2-digit numbers, they were followed by the other numbers. (Sam)

John's response was typical among those who made errors. He wrote:

> The answer is: 86 + 52 = 138, I wrote the largest number I could find using the given digits (86), then I wrote the largest number out of the remaining digits (52) in this way I found the largest sum.

Paul, who also concluded that the result was 138, reasoned that:

> The answer is: 86 + 52 = 138. I ordered the numbers from the largest to the smallest, then I grouped the digits into 2-digit numbers.

Ed, on the other hand, justified his response saying:

> My answer is 68 + 52 = 120, I looked for a sum which gives a 3-digit number.

While Terry explained:

> I tried out a lot of calculations and I found that 85 + 68 = 151.

From the responses we received we see that the pupils who answered correctly took into account the place-value system, although not all of them explained their reasoning clearly. The pupils who made errors used more idiosyncratic strategies, like finding the biggest numbers without referring to the principles of the decimal system (John and Paul). In contrast, Terry and Ed actually tried to give an answer without taking into account all of the task specifications.

Pause for thought

As a teacher, how might you work with pupils who gave such a range of answers?

Discussion with the teachers

We discussed the issue with the teachers in our project group and they suggested that, after working through the problem with the children, they could extend the task in a variety of different ways and work on them with the pupils. When they tried such an approach they found it to be an extremely useful way into understanding pupils' strategies when faced with such problems. This in turn helped them develop new, easy to administer, problems to enhance their understanding of pupils' – often subtle – misconceptions of place value while using it in a complex task. The new tasks the teachers suggested are shown in Figure 2.3.

Figure 2.3 Addition tasks to explore children's understanding of place value

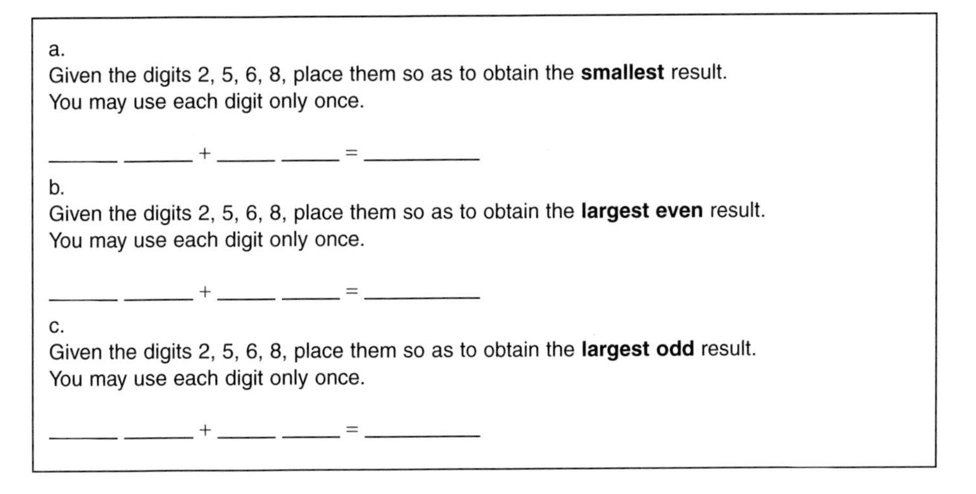
a.
Given the digits 2, 5, 6, 8, place them so as to obtain the **smallest** result.
You may use each digit only once.

___ ___ + ___ ___ = ______

b.
Given the digits 2, 5, 6, 8, place them so as to obtain the **largest even** result.
You may use each digit only once.

___ ___ + ___ ___ = ______

c.
Given the digits 2, 5, 6, 8, place them so as to obtain the **largest odd** result.
You may use each digit only once.

___ ___ + ___ ___ = ______

The teachers also suggested modifying the task to such an extent that there is no solution (see Figure 2.4) which, they discovered, proved to be a highly motivating exercise. Try it!

Figure 2.4 Addition task to deepen the pupils' understanding

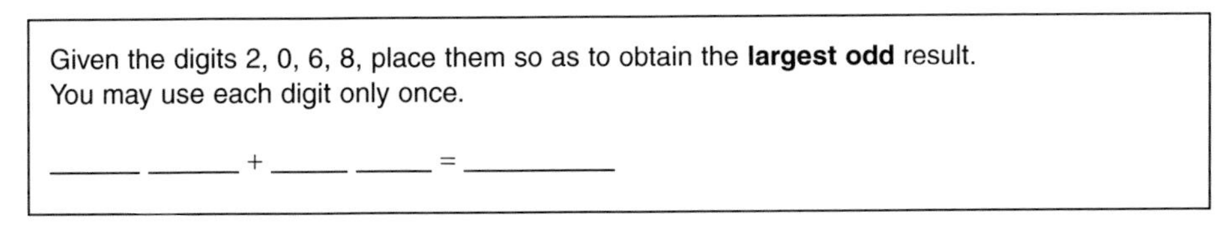
Given the digits 2, 0, 6, 8, place them so as to obtain the **largest odd** result.
You may use each digit only once.

___ ___ + ___ ___ = ______

Moreover, the teachers noted that pupils quickly learnt that they must combine reasoning with mathematical principles.

They also observed an added, and unexpected, bonus in that some children began to spot certain facets of the numbers they were working with: for example, that using even numbers they can only create two-digit even numbers, and that the sum of even numbers is always even. This reasoning led the teachers to develop another new task (see Figure 2.5). Again you might like to try it. You might even discuss it with your colleagues and consider how you might build on our initial ideas.

Figure 2.5 New task developed by teachers

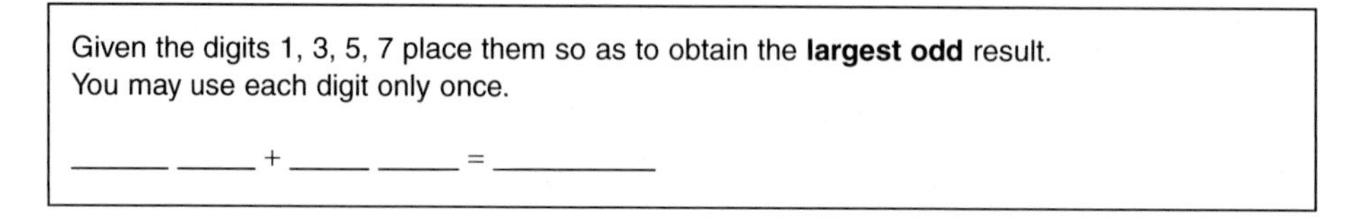
Given the digits 1, 3, 5, 7 place them so as to obtain the **largest odd** result.
You may use each digit only once.

_____ _____ + _____ _____ = _________

If you are mathematically minded you might appreciate the concluding summary (presented in Figure 2.6) that attempts to explain 'what is going on behind the scenes' of the above. Otherwise we suggest you skip to the next section.

Figure 2.6 Mathematical reasoning for the addition task

Given the digits a, b, c, and d,

The 2-digit numbers we can create are: $a \times 10 + b$, and $c \times 10 + d$

We can write these as:

$10a + b$ and $10c + d$

In order to find out the biggest sum we have to find out the maximum of

$10a + b + 10c + d = 10 (a + c) + 1 (b + d)$

We can see that a and c are interchangeable, as are b and d.

To maximise the sum, a and c need to be the largest numbers.

The subtraction task

How did you and, if applicable, your class get on with this task (Figure 2.7)? How do your results compare with those we present below? Most of the project

Figure 2.7 Arranging four digits to create a calculation involving subtraction

Given the digits 2, 5, 6, 8, place them so as to obtain the largest result.
You may use each digit only once.

_____ _____ − _____ _____ = _________

children (aged 10–11 years) answered correctly and gave similar explanations such as:

> To begin with, I chose the largest number, and at the end the smallest.
>
> To obtain the largest result in subtraction I have to subtract the smallest number from the largest.

Some children – such as Sam – wrote the right solution but gave only a partial explanation:

> I have to put in biggest number in the place of the tens digit of the first number and the smallest number in the tens digit of the second number.

In contrast, others gave the right explanation and put the numbers in the correct place but were unable to complete the calculation correctly, producing answers such as 86 – 25 = 41 and 86 – 25 = 66. Still others wrote the right numbers, completed the calculation accurately, but gave very weak explanations such as 'That is what I succeeded to find'.

Pause for thought

What do these results tell you about what the children know? What else – if anything – would it be helpful for you to know? Why?

As we described earlier, few pupils in any of the age groups we worked with gave incorrect solutions. For example:

> I built the smallest numbers and I wrote 28 – 25 = 3. (Julie, aged 10)
>
> When I subtract a big number from a big number I get the smallest difference: 85 – 62 = 23. (Jack)

Melanie produced the same calculation as Jack, but reasoned:

> I followed my previous work and I built the biggest numbers.

Jacob explained:

> The answer is 86 – 52 = 34. I have to put the biggest numbers in the tens place to get the biggest result.

Unfortunately we were unable to interview him as it would have been interesting to ask Jacob to explain the difference between the subtraction and addition problems.

Phoebe wrote:

> 68 – 25 = 43 I made the smallest numbers I could to get the biggest difference.

Interestingly their errors seemed to be very individual and we could detect no pattern to them. Some of the pupils changed the task: instead of finding the largest result they tried to find the smallest (Julie and Jack). Some changed the conditions and used a digit more the once (Julie). Some used explicit place-value considerations, but unfortunately did not reason precisely: Melanie got the correct answer but did not present the whole explanation, while Jacob did not reason well or find the correct solution.

Discussion with the teachers

Did you and/or your class find the addition task more difficult than the subtraction one? In trying to understand the phenomenon, one of the project teachers, Dianna, said in the discussion:

> The children take into account fewer mathematical considerations in the subtraction problem than in the addition problem and they only focus on the size of the numbers and not on place value. If they have a good understanding of the operations, they know that to obtain the largest result for subtraction they have to subtract the smallest number from the biggest number. It is not exactly the same, but it is similar to tasks like: 'If X is the unknown number, place "<", ">", or "=" in □ in the following equation: $X - 2 \square X - 10$.'

As they considered this, the teachers concluded that in such examples you do not have to know what X is but you do need to be able to appreciate that the expression $X - 2$ is larger than $X - 10$ because, as Jean explained, 'The less you remove the more you have'. The teachers were then given the following challenge:

Challenge

Devise tasks which might provide further insight into pupils' knowledge of the role of place value in subtraction.

The teachers' responses are given in Figure 2.8.

Thinking back to Figure 2.4 – the task where there was no solution – Jean, one of the teachers, raised the question

> Can we find a similar task for which there is no answer? I find such examples very helpful when trying to understand the limits of different concepts.

Challenge

Can you find such a task?

Figure 2.8 Subtraction tasks to explore children's understanding of place value

a.
Given the digits 2, 5, 6, 8, place them so as to obtain the smallest result.
You may use each digit only once.

_____ _____ – _____ _____ = ________

b.
Given the digits 2, 5, 6, 8, place them so as to obtain the **largest even** result.
You may use each digit only once.

_____ _____ – _____ _____ = ________

c.
Given the digits 2, 5, 6, 8, place them so as to obtain the **largest odd** result.
You may use each digit only once.

_____ _____ – _____ _____ = ________

One of the more mathematically minded of the group suggested:

> Let us try to set out the mathematical reasoning behind the original task as we did for the addition problem.

The result is given in Figure 2.9, but feel free to skip it if you are feeling mathematically squeamish and move on to the division task.

Figure 2.9 Mathematical reasoning for the subtraction task

Given the digits a, b, c, and d the 2-digit numbers we can create are:
$a \times 10 + b$ and $c \times 10 + d$

We can write these as:

$10a + b$ and $10c + d$

In order to find the biggest difference you have to find the maximum value possible for,
$10a + b - (10c + d) = 10 (a - c) + 1 (b - d)$.

To do this you need to maximise $a - c$ which means:

a will be the largest of the given numbers

c will be the smallest of the given numbers

From the remaining numbers you need to maximise $b - d$.

Thus b will be the largest remaining number and

d will be the smallest.

The division task

As we discussed earlier, in our experience children found the division task (Figure 2.10) easier than the multiplication. Did you? Did your pupils? As we

Figure 2.10 Arranging four digits to create a calculation involving division

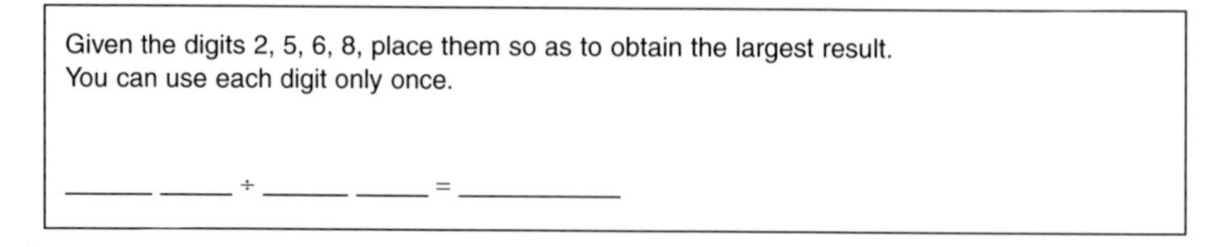
Given the digits 2, 5, 6, 8, place them so as to obtain the largest result.
You can use each digit only once.

___ ___ ÷ ___ ___ = ______

present our findings from both tasks, we suggest you compare and contrast the results with your own. At the end of the chapter we will discuss the teachers' reactions to both tasks so you can see how they compare with your own.

Pause for thought

How would you expect students to answer the division task?

Interestingly we found responses to the division task similar to those in the subtraction task. Most of the pupils presented correct answers, but they differed in their reasoning.

> I have to begin with the largest number and then divide it by the smallest one. (Sally)
>
> When I divide by the smallest number I obtain the largest result. (George)
>
> To have the largest result I must find the smallest number which goes into the large number most times. (Toby)

Some individuals observed that the largest result is not a whole number:

> I divided by 25 as many times as I could and I was left with a remainder of 11.

Contrary to the subtraction task, we found some pupils presented correct answers but did not explain how they obtained them. Others, such as George above, only produced a partial, albeit correct, explanation. There were also some children who did not answer the problem at all.

We found three pupils whose reasoning was partially correct, but as they were not accurate, their answers were incorrect. For example, Peter wrote:

> $56 \div 28 = 2$ because I must place the biggest digits in the first number – the dividend – and the smallest in the second number – the divisor.

Notice that he differentiated between digits and numbers but that his answer is only partially correct and eventually he presented a wrong answer. It may be that – in contrast to the other tasks which only involved whole numbers – the three children were put off by fractional results. Perhaps they tried to 'play' with the numbers in order to find a whole number in the result while forgetting their initial reasoning and, as a result, acted differently than they had originally intended. In other words, they did not use the largest digit for the dividend as they said they would. Nor did they take into account the place value of the two-digit number to obtain the largest number from the two larger digits. Might there be other explanations?

There was another group of pupils who wrote '86 ÷ 52 = 34 (remainder 1)', but they each gave different explanations such as:

> After a lot of calculations that is the best result I found. (Mandy)
>
> I answered all the previous tasks the same way. (Freddy)

(This interesting explanation leads us to another question about consistency of students' reactions – when and why?)

Some of the other children, such as Suzanne, simply wrote the mathematical expression without any explanation, adding:

> That is the answer.

Finally, as with previous tasks, some individuals did not consider all the constraints of the task and gave responses such as '86 ÷ 2 = 43', '86 ÷ 62' and '67 ÷ 17'. What might we deduce from such answers? How could we ascertain whether our conclusions were correct?

The multiplication task

Pause for thought

If you do not have a class of your own, how might you expect children to respond to the multiplication task?

Figure 2.11 Arranging four digits to create a calculation involving multiplication

Given the digits 2, 5, 6, 8, place them so as to obtain the largest result.
You may use each digit only once.

____ ____ × ____ ____ = ________

The multiplication task (Figure 2.11) proved to be the most surprising for our teachers. Few children solved it successfully. The same phenomenon was observed in all the project classes. Sometimes pupils who found the correct numerical answer did not give a full explanation. For example:

> The largest digits gave the largest result. (Sandra)
>
> We have to put the largest digits in the tens value and to check in which order to place the units. (John)
>
> We have to multiply big numbers by big numbers. (Stewart)

The most typical error we found was $86 \times 52 = 4472$, but we received different explanations such as:

> I multiplied the largest numbers.
>
> I multiplied the largest number by the smaller.
>
> This is the answer I got after my calculations.

Another common error was of the type given by Jude:

> $82 \times 56 = 4592$. In this way I arrived to the correct answer.

Some pupils gave the same numerical answer without any explanation.

While analysing this problem there are several mathematical considerations to take into account, with place value being one of the most important. The other is reasoning which number to multiply by which so as to obtain the largest result: while it is relatively 'easy' to see that you need put the largest digits in the tens place, to multiply tens by tens to get hundreds, it is much harder to plan in which order to place the units digits.

Discussion with the teachers

Pause for thought

Did you – like our teacher colleagues – find the division task easier than the multiplication one? If not, why do you think there was a difference? Do you think it was because of your mathematical background or more to do with your understanding of the tasks?

Looking at the children's results, the project teachers took little time to discuss the division task as they felt it was relatively easy to solve and did not yield much information about the children's mathematical understanding. They were, however, highly intrigued by the multiplication task and quickly became fairly mathematical in their discussion – but try not to be put off! Jean opened the discussion as follows:

> When we were pondering about two possibilities for the multiplication task, we immediately spotted that the tens digit must be the larger – 8 and 6 in this task, it means $[(8 \times 10) + a]$ multiplied by $[(6 \times 10) + b]$ – but I hesitated for a short while among the units digit (a,b), e.g. between 82×65 and between 85×62.

The conversation continued:

> Soffia: We can calculate it.
> Dianna: You are right, but let us try to think more generally before working it out. At the end we will check our decision with some calculations. If the larger digit in the tens place (8) is in the multiplicand (the first factor),

I am sure about 82 × 65. You see we have to choose among 2 or 5 for the units place which means that we have to choose between multiplying 8 tens (80) by 2 or by 5 – it is clear that 5 is preferable.

Soffia: We end up with 82 × 65.

Jean: This seems a good way to help our pupils, but first we have to convince them about the tens digit.

Soffia: That's quite easy because we can use estimation. 80 × 60 is 4800 which is higher than any of the other possibilities such as 80 × 50 or – and I don't believe any child would try it – 80 × 20.

Dianna then suggested they produce an abstract account of the procedure (see Figure 2.12) similar to those in Figures 2.6 and 2.9. Again, feel free to skip this if you wish.

Figure 2.12 Mathematical reasoning for the multiplication task

> Given the digits a, b, c, and d the 2-digit numbers we can create are:
> $a \times 10 + b$ and $c \times 10 + d$
>
> We can write these as:
>
> $10a + b$ and $10c + d$
>
> If the first factor is '$10a + b$' and the second factor is '$10c + d$,' we have to find the biggest product of: $(10a + b) \times (10c + d)$ which means we have to find the maximum of $100ac + 10ad + 10bc + bd$
>
> From this expression it is clear that a,c must be the largest digits. And our hesitation about the digits d and c is clear. The expression 'bd' will be the same either way but we have to decide about '$10ad + 10bc$'. I think that if a is bigger than b then it is obvious that if we multiply the biggest numbers we will have the biggest product.

The teachers then wanted to move into broadening the tasks so that they could deepen their understanding of pupils' knowledge. Interestingly, almost certainly as a result of their earlier discussions, they quickly came up with some examples similar to the division problem.

Challenge

Can you find such a task?

Seeing these examples (Figure 2.13) Dianna announced: 'It is the same problem as we had before.' The other teachers then wondered whether it might be more interesting to give children the problem shown in Figure 2.14 and challenge them to explain why it is impossible.

The discussion ended with Jean proposing:

> We could continue asking them – how do we have to change the given numbers so that we can find an answer to the problem?

Figure 2.13 Division and multiplication tasks to explore children's understanding of place value

a.
Given the digits 2, 5, 6, 8, place them so as to obtain the **smallest** result.
You may use each digit only once.

____ ____ ÷ ____ ____ = ______

b.
Given the digits 2, 5, 6, 8, place them so as to make **the result a whole number.**
You may use each digit only once.
____ ____ ÷ ____ ____ = ______

And for the multiplication task...

c.
Given the digits 2, 5, 6, 8, place them so as to obtain the **smallest** result.
You may use each digit only once.

____ ____ × ____ ____ = ______

d.
Given the digits 2, 5, 6, 8, place them so as to obtain the **largest even** result.
You may use each digit only once.

____ ____ × ____ ____ = ______

Figure 2.14 Arranging four digits to create a calculation involving multiplication – the case of impossible solution

Given the digits 2, 5, 6, 8, place them so as to obtain the **largest odd** result.
You may use each digit only once.

____ ____ × ____ ____ = ______

Concluding remarks

For some of you the approaches used in this chapter may have been unfamiliar, making it quite hard work but – we hope – fascinating and informative. To summarise, we have discussed pupils' understanding and misunderstanding (or misconceptions), focusing on two important aspects of mathematics: place value (which is one of the basic concepts in the number system we use) and the meaning of the four basic operations. In the tasks we presented, children had to combine the two aspects in order to produce a correct answer. To successfully complete the tasks pupils had to have a sound understanding of place value and the operation – for example, addition – in which they were engaged. By asking them to explain the thinking behind their responses we were able to unravel where any misconceptions they had might lie. We shall return to these tasks in Chapter 6, where we will explore them from another perspective.

And finally, having read the chapter and tried some of the examples, perhaps those of you who swerved around them might now feel bold enough to tackle Figures 2.6, 2.9 and 2.12. They are not unduly hard for those who enjoy algebra but we can appreciate that, to the uninitiated, they might look rather daunting at first sight.

Further reading

Fuson, K.C. (1992) Research on whole number addition and subtraction. In D.A. Grouws (ed.), *Handbook of Research on Mathematics Teaching and Learning* (pp. 243–275). New York: Macmillan.

As children shift from mental calculations to using calculators etc. there is a huge need to understand a variety of different conceptual structures that result in different solution procedures for some problems. This chapter explores such issues, from starting with the early use of numbers by young children up to understanding addition and subtraction based on place value.

Greer, B. (1992) Multiplication and division as models of situations. In D.A. Grouws (ed.), *Handbook of Research on Mathematics Teaching and Learning* (pp. 276–295) New York: Macmillan.

The chapter begins with the variety of situations related to multiplication and division, followed by representations of the operations. The discussion then extends beyond positive whole numbers. The chapter ends with several suggestions for teaching these operations.

Peled, I. and Segalis, B. (2005) It's not too late to conceptualize: Constructing a generalized subtraction schema by abstracting and connecting procedures. *Mathematical Thinking & Learning,* 7(3), 207–230.

In this study children (year 6) were encouraged to make connections between subtraction procedures in whole numbers, fractions, and decimals. They received mapping instructions that encouraged them to generalise procedural steps. This process led to the abstraction of mathematical principles underlying these procedures. For example, students realised that regrouping and place value relations are involved, and that regrouping does not change the total amount of the number.

Insights into children's intuitions of addition, subtraction, multiplication and division

Dina Tirosh, Pessia Tsamir and Sara Hershkovitz

In the previous chapter we discussed the concepts of addition, subtraction, multiplication and division. How do you think 11-year-olds would reply if you asked them to define them? Whenever we ask the question children almost invariably instantly respond – and with great confidence – along the lines of, 'addition makes bigger', 'subtraction makes smaller', 'multiplication makes bigger' and 'division always makes smaller'.

Interestingly, when they stop to think about it, however, the very same children have sufficient mathematical knowledge to deduce that such statements are not always true. Most of them know, for instance, that adding a zero to a number or subtracting zero from a number does not 'change' the number. Some are also aware of the fact that zero times a number is zero and that dividing a number by one does not change the number. What is of particular interest here though is that, despite this knowledge, their intuitive (immediate and confident) responses to the question are as given above.

In this chapter we will describe some of the intuitive – and yet sometimes incorrect – beliefs that children often hold about addition, subtraction, multiplication and division. We will consider their likely origins and then suggest some of the implications for classroom practice. We will accompany Debby, an experienced teacher of upper primary classes. She leads a group of 12 teachers who meet once a month to discuss various issues regarding their teaching. Typically Debby begins the meetings by presenting some children's work for her colleagues to consider. From time to time she will mention research findings if she thinks they have anything relevant to add to the discussion. Frequently she refers to the work of Fischbein (1987, 1993; Fischbein *et al.*, 1985), a psychologist whose work mainly focused on the role of intuition in mathematics and science education. As you read the chapter, we invite you to act as an invisible participant in three such sessions.

Debby's group of teachers

Session 1: Introduction, addition and subtraction

Debby began the session by presenting an exercise that she had recently given to her class of 11-year-olds. The assignment included two types of tasks: 'calculations' (Figure 3.1) and 'true or false statements' (Figure 3.2). She asked the participants in the workshop to start off by responding to each task.

Figure 3.1 Debby's calculation task

Task I. Please calculate the following:

1) $5.14 + 0$	6) $20 \div 5$
2) $7.3 - ? = 7.3$	7) $8 \div 0.25$
3) 0.14×0.3	8) $20 \div 5$
4) 5×0.25	9) $5 \div 20$
5) $4.04 \div 4$	10) $0.25 \div 8$

Challenge

Take a few moments to try the tasks for yourself. How do your responses compare to those of the teachers in Debby's group?

Figure 3.2 Debby's true or false statements

Task 2. 'True or false'?

Read carefully each of the following statements. Then decide whether it is true or false. Give reasons for your answers.

1) In an addition problem, the sum is always greater than each addend.
True/False: Why? ______________________
2) In a subtraction problem, the difference is always smaller than the minuend
True/False: Why? ______________________
3) In a multiplication problem, the product is always greater than either factor.
True/False: Why? ______________________
4) In a division problem, the divisor can be larger than the dividend
True/False: Why? ______________________
5) In a division problem, the quotient must be less than the dividend
True/False: Why? ______________________
6) In a division problem, the divisor must be a whole number.
True/False: Why? ______________________

Figure 3.3 A sample of Ben's response

Task 1. Compute: 5.14 + 0

Ben wrote: 5.14 + 0 = 5.14

Task 2. 'True or false'? Explain why.

1) In an addition problem, the sum is always greater than each addend.

Ben wrote: True

The teachers concluded that all but one of the statements (Statement 4) in Task 2 are false. Debby then explained that, having given the tasks to her class, she decided to mark Ben's work first as he was one of her best mathematicians and she was confident that he would answer everything correctly. She was, however, surprised and frustrated when she realised that this was not the case (Figure 3.3). She then focused on the work of all the students, addressing each operation in turn. Although she had read about the phenomena she observed in the children's work in several articles, she still felt disappointed.

Challenge

Why do you think Ben responded in the way he did?

Based on her experience and on her recent reading, Debby produced some follow-up tasks she had given her class to help her explore what might have been going on. She started with addition.

Addition

Debby pointed out to the group that, while Ben provided a 'true' response to the addition statement in Task 2, he had argued, incorrectly, that in an addition problem the sum is always greater than each addend. She adjusted the task therefore and asked her class to provide explanations for their answers. Ben wrote:

> I'm thinking about two friends. One has 12 cookies in his bag; the other has 14 cookies in his bag. They decided to put their cookies on the table. ... [t]here are more cookies on the table than in each bag.

Debby then explained to the group that 28 of the 30 pupils in her class had also given a 'true' response, so she analysed their responses and categorised them into various types, in accordance with the related research (e.g. Fuson, 2003). She ended up with the list shown in Figure 3.4. There was considerable

Figure 3.4 Children's correct and incorrect responses to 'addition makes bigger'

Response – True (28 pupils)
Explanations

1. ***Dynamic situation (adding to)***
   Sally: *Sally had 4 apples. Dan gave her 5 apples. She now has 9 apples (9 is bigger than 5 and bigger than 4).*
   *(This type of addition word problem was given by **10** pupils)*
2. ***Static situation (putting together)***
   Ben: *I'm thinking about two friends. One has 12 cookies in his bag; the other has 14 cookies in his bag. They decided to put their cookies on the table. There are 26 cookies on the table. So, there are more cookies on the table than in each bag.*
   *(This type of addition word problem was given by **6** pupils)*
   *(Similar word problems were written by **4** pupils).*
3. ***Adding is combining disjoint sets***
   Tamara: *Addition means putting together. When you add two things, you come up with something that is greater than each of them, because you put things together...*
   *(Similar explanations were given by **4** pupils).*
4. ***Explicit beliefs about addition***
   Judy: *Addition always makes bigger.*
   *(Similar explanations were given by **8** pupils).*
5. ***Zero as an exceptional***
   John: *The answer that I get when I add two numbers is almost always bigger than the numbers that I add. When I add zero to a number I get the same number. So zero is exceptional, and except for zero it is so. So – it is TRUE.*
   *(Similar explanations was written by **2** pupils).*

Response – False (2 pupils)
Explanations

6. ***Zero refutes the statement***
   Dora: *Task 1 shows that 5.14 + 0 = 5.14. I ended up with the first number. This shows that in an addition problem, the sum is not always greater than each addend.*
7. ***The negative numbers refute the statement***
   Cindy: *My mother taught me about negative numbers. I know that when you add a negative number to a number you get a number that is smaller than the number that you had.*

discussion among the teachers about each type of response given. Below are the main comments which were made:

- The first four types of explanation are related to each other. The first two describe two major situations that often come to mind when thinking about addition. In these situations, two independent groups of items (disjoint sets) are united. This general model is described in explanation 3. In the intuitive representation of this model, each of the two sets has at least one element (i.e. none of these sets are empty). Thus, it is only to be expected that the intuitive model of addition is that of putting together two disjoint sets of objects. Consequently, we tend, intuitively, to view addition as 'making more' (explanation 4). Our life experience strongly supports the intuitive view that addition 'makes bigger' as we frequently add one quantity to another and end up with a larger amount. Furthermore, it seems that the response to statement 1 in Task 2 is based on a solid, intuitive belief about addition. The conflicting result of the calculation ($5.14 + 0 = 5.14$) is probably not considered when children are offering an intuitive response.
- Responses 5 and 6 were discussed together. The teachers commented that such explanations relate to grasping the unique nature of proofs and refutations of mathematics. Both responses addressed the case of zero. In response 6, Dora correctly concluded that this case violates the statement. Interestingly John (explanation 5) had adopted a similar approach but mistakenly argued that exceptions are permitted in such a (universal) statement in mathematics.
- Explanation 7 led to a discussion of a situation that the teachers in the working group commonly encountered – parents had taught their children topics in mathematics before these topics are taught in schools.

During the discussions, the teachers suggested several teaching strategies which might be helpful in addressing the emerging misconceptions. Debby clarified whether these strategies were specific to addition. The teachers concluded that they were not so Debby suggested that they discuss the responses to the other operations first, and then consider some of the teaching implications.

Subtraction

Debby distributed the task in Figure 3.5 amongst the teachers, to stimulate discussion.

Figure 3.5 Debby's subtraction task

Subtraction

Imagine asking some 11-year-olds to respond to the statement:

In a subtraction problem, the difference is always smaller than the minuend.
True/False: Why? ____________________

Please try to list different types of responses that pupils at this age are likely to give.

Figure 3.6 Correct and incorrect responses to 'subtraction makes smaller'

Response – True
Explanations

1. ***Dynamic situation*** *(take away)*
   Sally had 5 apples. She gave 2 apples to Dan. She now has 3 apples (5>3).
2. ***Dynamic situation*** *(difference/building up/missing addend)*
   I want to buy a shirt that costs £10. I have £3. How many more pounds do I need to buy the shirt?
3. ***Static situation*** *(difference-comparison)*
   Sue has eight apples. Ann has four. How many more apples does Sue have than Ann?
4. ***Explicit beliefs about subtraction***
   Subtraction always makes smaller.
5. ***Zero as an exception*** *A number minus a number is always smaller than the first number, except for minus zero (10 – 0 = 10, 9 – 0 = 9 and so on), but zero is an exception.*

Response – False
Explanations

6. ***The case of zero refutes the statement***
   10 – 0 = 10, I ended up with the first number. This shows that subtraction does not always make the result smaller.
7. ***The negative numbers refute the statement***
   I know that when I subtract a negative number from a number I get a number that is larger than the first number.

The teachers worked in three groups. They collated their discussions into a list (Figure 3.6) based on their knowledge of the different types of subtraction. These included distinguishing between problems which ask you to change, combine, and compare (Fuson, 2003) and the five types of subtraction outlined by Haylock and Cockburn (2003). They also discussed the importance of the place of the unknown in the problem (e.g. $5 - 2 = \square$ is the easiest case, $5 - \square = 2$ tends to be harder and $\square - 2 = 3$ is usually even more demanding).

Having completed the task the teachers were curious to know how well their responses matched those of Debby's class. Debby said that the children's most common explanation was the first one given by the teachers, while some gave explanations similar to the fourth and fifth and one pupil gave the sixth. Explanations 2, 3 and 7 were not given by any of 11-year-olds.

Pause for thought

How might you explain the range of responses given by Debby's class? What teaching strategies might you suggest to address any issues raised?

The children's responses surprised some teachers in the working group. Debby, too, was slightly taken aback when she realised that almost all of her class argued that subtraction makes smaller and that none of the 11-year-olds considered subtraction in terms of the difference between two quantities. At this point she recalled research articles suggesting that the intuitive model of subtraction is 'take away'. Fischbein (1993: 238), she explained, provided an example of the intuitive model of subtraction: 'You have in a container a number *A* of objects (for instance, marbles) and you want to take out a number of them, *B*'. She added that Fischbein also concluded that other interpretations of subtraction are more demanding. The 'difference by building up' interpretation, for instance, intuitively suggests addition. The 'taking away' intuitive model encourages the belief that 'subtraction makes smaller' and consequently we tend, intuitively, to view subtraction as 'making fewer'. With these words the first session came to an end. The teachers commented that multiplication and division are even more complicated. Debby promised to deal with these operations in the second session.

Session 2: Multiplication and division

Debby started the second session by explaining that the tasks that they were going to work on had already been done by children in various countries (e.g. the UK, Italy and Israel). She also pointed out that, regardless of country, the incorrect responses were similar.

Multiplication

Debby began by presenting two problems (Figure 3.7) that were included in Fischbein's study on intuitive models. She asked the teachers to list the incorrect responses they would predict for each problem. What do *you* think they might be?

Figure 3.7 The multiplication task the teachers were set

The task, for the pupils, was:

For each problem, write the number sentence you would use to solve it (please *do not* complete the calculation!).

1. From 1 kilo of wheat, you get 0.75 kilos of flour. How much flour do you get from 15 kilos of wheat? *The number sentence*: ______________
2. One kilo of a detergent is used in making 15 kilos of soap. How much soap can be made from 0.75 kilos of detergent? *The number sentence*: ______________

Your task: Write the incorrect responses you would predict for each problem. Describe possible reasons for these responses.

Pause for thought

Which of the two tasks do you think might be the more demanding? Why?

The teachers in Debby's group thought that the second problem was more demanding than the first which, interestingly, proved to be the case when 628 upper primary/lower secondary (i.e. 10–15 year-olds) were tested in Pisa, Italy (Fischbein *et al.*, 1985). In that study the percentages of correct answers were: for problem 1, 79% (age 11), 74% (age 13) and 76% (age 15); and for problem 2, 27% (age 11), 18% (age 13) and 35% (age 15).

So, if the solution to both is 15×0.75, why is one harder than the other? Debby explained to the group that, in the first problem, the first factor is a whole number (15), whilst in the second, it is a decimal. She commented, however that, although this should not have made any difference (because multiplication is a commutative operation), Fischbein pointed out that the intuitive model of multiplication is repeated addition. Thus, 15 times 0.75 has an intuitive meaning, but try representing 0.75 lots of 15 using repeated addition! Fischbein (1993: 237) concluded:

> In a multiplication $A \times B$ ... If A is a decimal, the student will not directly grasp the solving procedures intuitively. The 'repeated addition model', operating behind the scenes, will prevent the right solution instead of facilitating it. As an effect of this situation ... the student is led to believe, intuitively, that 'multiplication makes bigger'.

As a consequence of the above, pupils tend to write a division expression for the second word problem, assuming that this would lead to a 'smaller number'.

Debby reported that about half of the children in her class incorrectly argued – in accordance with their intuitive beliefs about multiplication – that the statement 'In a multiplication problem, the product is always greater than either factor' is true. She ended the discussion on multiplication by presenting the teachers with the following challenge:

Challenge

Ask your pupils to undertake the two multiplication word problems listed above.

Try to analyse their responses. Do they agree with the behaviour suggested by the 'intuitive model' theory?

Division

For division, Debby presented the teachers with the task set out in Figure 3.8. Try it!

Figure 3.8 Debby's division task for teachers

Division

Please write two different word problems that would be solved by using the following number sentences (try to use no more than two contexts):

1) $16 \div 2$ 2) $2 \div 16$ 3) $1.25 \div 5$ 4) $5 \div 0.25$

Which of these tasks do you think would be the most difficult for 11-year-olds?

The teachers worked individually. After a while Dan, one of the teachers, volunteered to present his problems, involving two models of division. Using the equal-sharing model he wrote three problems on the board, using friends and cheese:

1) 16 ÷ 2 Two friends bought 16 kilograms of cheese. If the cheese was equally shared, how much did each get?
2) 2 ÷ 16 Sixteen friends bought 2 kilograms of cheese. If the cheese was equally shared, how much did each get?
3) 1.25 ÷ 5 Five friends bought 1.25 kilograms of cheese. If the cheese was equally shared, how much did each get?

Dan explained that he did not write a 'sharing' problem for 5 ÷ 0.25 because '0.25 is not a whole number' and so he used the measurement model instead, coming up with:

4) 5 ÷ 0.25 Peanuts are packed so that there are 0.25 kilograms in a box. How many boxes can be filled with 5 kilograms of peanuts?

The teachers agreed that 5 ÷ 0.25 was the hardest number sentence as it does not lend itself to the 'sharing' interpretation of division which is the type with which they are most familiar. Some teachers also concluded that the second task would be more demanding than the first and the third:

> Pupils will reverse the role of the divisor and the dividend, writing a problem for 16 ÷ 2 and not for 2 ÷ 16.

Figure 3.9 Common intuitive beliefs about division

The intuitive model of division is the sharing model. In this model an object or collection of objects is divided into a number of equal parts (the unknown is the number of the elements in each set). This model imposes three constraints on the operation of division:

1. The divisor must be a whole number greater than zero;
2. The divisor must be less than the dividend; and
3. The quotient must be less than the dividend.

Debby then presented the slide shown in Figure 3.9, explaining that she had created it on the basis of research articles that she had read on the topic (Fischbein, 1993; Tirosh, 2000). Asking the teachers to take the above into account, Debby then referred back to tasks presented in the first session (see Figure 3.2) and asked the teachers to complete Figure 3.10 estimating the percentage of children they thought would produce each response.

Pause for thought

What would your predictions be?

Figure 3.10 Debby's estimation task

Item in Figure 3.2	Statement	Incorrect response	% predicted
4	In a division problem, the divisor can be larger than the dividend	False	
5	In a division problem, the quotient must be less than the dividend	True	
6	In a division problem, the quotient must be less than the dividend	True	

The teachers worked in pairs and then the following conversation took place among the whole group:

Gili: In my class [10-year-olds], all or almost all the pupils will give these responses to these three statements.
Dori: In my class [9-year-olds] they will also ...
Avi: It's confusing. In my class [11-Year-olds] ... according to the sharing model, they will give these responses. But I'm almost sure that ... well, they solve so many problems like $5 \div 0.25$, or $2 \div 10$, so I feel that they will know that, at least they will know that statements 4 and 6 are false, ... they have seen so many such problems recently ... But the idea of the intuitive model – sharing – about statement 5 ... I just don't know...

Debby then reported on the unexpected percentages relating to incorrect responses she observed with her class of 11-year-olds: for statement 4, 56%; for statement 5, 87%; and for statement 6, 16%. Debby explained that most of the justifications for the *correct* responses from her pupils consisted of computational examples such as $2 \div 8$; $2 \div 0.5 = 4$, $4 \div 0.25$ for statements 4, 5 and 6, respectively. Apparently such responses to statements 4 and 6 were much more frequent than to statement 5. Why do you think this might be the case? If you stop to think about it, the examples for statements 4 and 6 could consist of mathematical open phrases without the 'answers'. In contrast, statement 5 addresses the relationship between the 'first number' and the 'answer', and thus both an open phrase and an 'answer' are needed.

Debby then became rather more technical in her discussion about Fischbein's work but, as you will discover, his conclusions provide some valuable insights into why children behave in the way they do in mathematics sessions. She explained that Fischbein (1993) had concluded that there was an interaction between *algorithmic, formal, and intuitive* components of mathematical activities:

- The *algorithmic* component consisted of both the knowledge of how to carry out a finite sequence of steps from a given mathematical problem to its solution and being able to provide a valid mathematical explanation for each

step (e.g. knowing the standard procedure for multiplication and being able to explain each step).
- The *formal* component refers to axioms, definitions, theorems and proofs. It includes knowledge of what is 'right', 'permitted' and 'acceptable' versus what is 'wrong', 'forbidden' and 'unacceptable' in mathematics (e.g. knowing that the sum of any two odd numbers is even and being able to prove it).
- The *intuitive* component is a kind of knowledge which is accepted immediately and confidently without a feeling that any kind of justification is required.

Debby pointed out that in the first years of schooling addition, subtraction, multiplication and division are discussed, almost solely, with counting numbers. Later on, children's knowledge of numbers is expanded to include fractions and decimals and this enlargement reveals many misconceptions about the operations (see, for instance, Kilpatrick *et al.*, 2003; Lester, 2007). Debby commented that Fischbein argued that many of these difficulties result from intuitive beliefs about the operations – beliefs that hold for the counting numbers but not necessarily for other types of numbers.

To illustrate the above in a practical context, Debby referred to Figure 3.11 which demonstrates the application of the framework to analysing mistakes children often make when dividing decimals. The teachers then started talking about ways of overcoming the errors and the inconsistencies that they had previously discussed. Debby said that these issues would be discussed in the third, and final, meeting.

Figure 3.11 Dividing decimals: algorithmic, formal and intuitive mistakes

- *Algorithmically*-based mistakes: Various 'bugs' in computing division expressions (e.g. Task 1- Part 4, a common, incorrect response to 4.04 ÷ 4 is 1.1. The 0 in 4.04 is ignored).
- *Formally*-based mistakes: Incorrect performance due to either limited conceptions of the notion of decimals or inadequate knowledge of the properties of the operations (e.g. 'division is commutative, therefore, 5 ÷ 20 = 20 ÷ 5 = 4').
- *Intuitively*-based mistakes: Incorrect responses rooted in adherence to the intuitive model of division (e.g. '5 ÷ 20 is impossible because one cannot divide a small number by a larger number because it is impossible to share less among more').

Session 3: Overcoming misconceptions – implications for teaching

Through conversations the teachers came up with two main approaches for overcoming misconceptions: teaching by analogy and the conflict teaching approach.

Teaching by analogy

In teaching by analogy, the session starts with pupils being presented with an 'anchoring task'. This task is designed to elicit a correct response. Later on, the pupils are presented with one or more essentially similar 'bridging tasks' in which a factor, or factors, that were identified as misleading, are progressively more evident. Finally, a 'target task', likely to elicit incorrect responses, is introduced. Such a sequence of instruction was proven to be effective in helping pupils overcome the impact of intuitive models (Clement, 1993; Stavy and Tirosh, 2000). Sometimes, the mere presentation of an anchoring task before the target task is sufficient to elicit correct responses to the target task, and there is no need for 'bridging task' (Tsamir, 2003).

Figure 3.12 Teaching by analogy: Dan's example

Stage 1: *The anchoring task* – One kilo of a detergent is used in making 15 kilos of soap. How much soap can be made from 6 kilos of detergent?
Stage 2: *The bridging task* – One kilo of a detergent is used in making 15 kilos of soap. How much soap can be made from 4.75 kilos of detergent?
Stage 3: *The target task* – One kilo of a detergent is used in making 15 kilos of soap. How much soap can be made from 0.75 kilos of detergent?

Using Figure 3.12, Dan demonstrated how he would use the teaching by analogy method to build up to one of the problems the teachers had discussed in their second session.

Pause for thought

Do you ever teach by analogy? If so, how? If not, is it a technique you might try?

The teachers immediately recognised Dan's method and several of them said that they often used variations of it. Avi called it the 'easy numbers' strategy (that is, replacing 'hard' numbers in a problem with smaller numbers which were easier to manipulate to assist in deciding on the appropriate operation). The teachers also suggested devoting time to forming 'new, adequate' connections between an estimation and the actual answer, especially when decimals are involved. Children need to appreciate that their initial estimation when calculating Dan's third-stage problem must be *less than* 15. In other words, they should be encouraged to conclude that division is not the way to solve the problem as 15 ÷ 0.75 is not smaller than 15. Rather they need to deduce that multiplication is the appropriate operation as 15 times 0.75 is less than 15.

Challenge

Can you think of other examples where teaching by analogy approach could be used to overcome the difficulties that are described in this chapter?

The conflict teaching approach

The second approach the teachers discussed involves cognitive conflict and first arose from the work of Jean Piaget. The three steps involved are outlined in Figure 3.13. Debby explained that Flavell (1977) claimed that cognitive conflict results in growth of understanding if the learner:

(a) notices the two conflicting elements;
(b) appreciates the inherent conflict in the two;
(c) searches for a resolution; and
(d) achieves a conceptualisation that resolves the conflict.

Figure 3.13 The three steps of the cognitive conflict teaching method

Step 1: *Exposing a misconception.* First, pupils are given a task likely to elicit an incorrect response.

Step 2: *Eliciting relevant, correct knowledge.* Pupils are presented with a situation related to the first task for which they are most likely to provide a correct response (e.g. a different representation of the same task and/or some concrete evidence and/or an extreme case).

Step 3: *Rising awareness of the contradiction.* Pupils are encouraged to examine their responses to the two previous stages and to spot the contradiction. Such a presentation often raises pupils' awareness of the inadequacy of their response to step 1, and thus leads to changing their initial incorrect response to a correct response.

She then demonstrated how she used the cognitive conflict method with her class when working on division. In step 1 (exposing the misconception) she asked the pupils to respond to the statement:

> In a division problem, the quotient must be less than the dividend.
>
> True/False: Why? ________________________

Debby noted that many of the children in her class incorrectly argued that this statement is true because 'division is sharing, and when you share, you and the other people get less than the total'. However, as most of the children in her class could correctly calculate expressions involving decimal numbers, she used this knowledge to plan step 2 (gaining relevant, correct knowledge). Debby asked the pupils to calculate the expressions

$5 \div 0.5$, $5 \div 0.25$, and $8 \div 0.1$

Practically all the pupils in her class provided correct responses to these three calculations so Debby was able to move on to the third step (rising awareness of the contradiction). She encouraged the pupils to examine their responses to step 1 and step 2, and to notice if there were any inconsistencies. The class then talked about the operation of division, considering not only whole numbers, but also decimals. As they were doing this the children noticed that their initial response ('division makes smaller') does not hold for 'all numbers' and thus the statement 'In a division problem, the quotient must be less than the dividend' is false.

Pause for thought

Imagine that you were asked to work with a class where 16% of the pupils had incorrectly agreed with the statement:

> In a division problem, the divisor must be a whole number.

1. Suggest a way of using the cognitive conflict approach to help pupils overcome this belief.
2. Are there other, possible occasions in which the cognitive conflict approach could be used to overcome the difficulties that are described in this chapter?

Finishing touches

By way of drawing the series of teachers' meeting to a close, Debby drew the group's attention to some more of Fischbein's (1987) views and they discussed the resulting practical implications. She emphasised that he urged teachers to take account of children's intuitions in the teaching process and advocated that the following should be included in any attempt to develop adequate intuitions:

- *Raising awareness of the role of intuition in the thinking processes.* Fischbein said that awareness of the nature and the impact of intuitions is a first step toward resisting them. He suggested discussing intuitive biases with children as a means of pointing out that they are natural and that there is a need to be aware of them, to identify their possible impact and to attempt to control this impact by attending to other sources of knowledge (algorithmic and formal).

In light of this recommendation, Debby and the participant teachers suggested giving children tasks that are likely to elicit intuitive, incorrect responses (e.g. Task 2), and then using the common, incorrect responses as a springboard to discuss the substantial impact of intuitive beliefs on pupils' responses to mathematical tasks.

- *Experiencing practical activities.* Fischbein concluded that intuitions cannot be modified by verbal explanations. He argued that in order to develop new 'adequate' intuitions it is essential to design specific, practical situations.

The teachers welcomed this recommendation with numerous suggestions including one from Avi relating to the statement: 'The quotient must be less than the dividend'. Avi proposed taking four apples into school and cutting each apple in half, to emphasise that the result to $4 \div 0.5 = 8$ half-apples. Such a demonstration would emphasise that the quotient is greater than the dividend.

'$4 \div 0.5 = 8$' or 8 half-apples

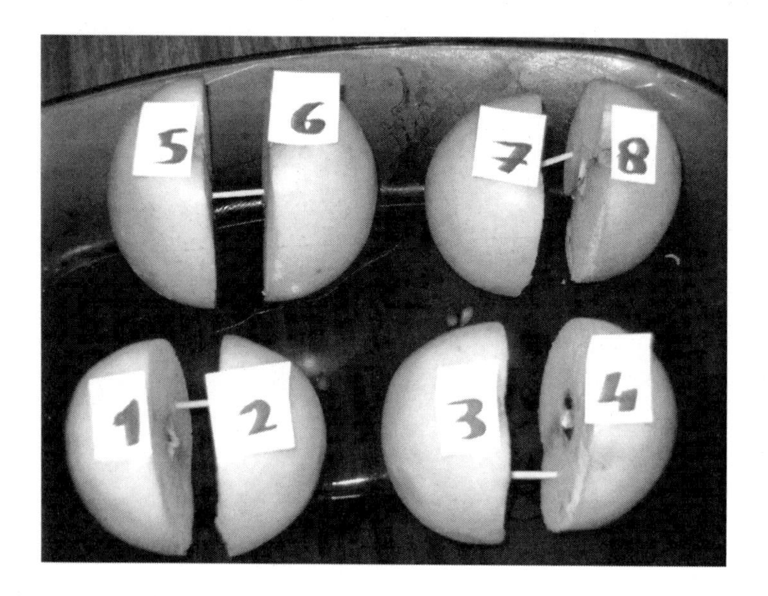

Pause for thought

Can you think of other simple, but equally compelling, examples?

- *Introducing the formal meaning of the concepts as early as possible.* Fischbein explained that there is an inherent didactical dilemma in the teaching and learning of mathematics. He argued that children's early experiences with numbers – both in and out of school – are mainly grounded in simple situations that are mathematically restricted to counting numbers. These initial intuitive interpretations often become very strongly attached to the concepts and, consequently, conceptual problems are caused when the numbers extend beyond those for counting, such as to decimals and negative numbers. Fischbein admitted that there is no general recipe for solving this dilemma and recommended starting, as early as possible, to prepare the child for understanding the formal, extended meanings of the concepts.

Challenge

How might Fischbein's ideas be incorporated into your school's current practice for teaching primary mathematics?

Further reading 

Fischbein, E. (1999) Intuitions and schemata in mathematical reasoning. *Educational Studies in Mathematics*, 38, 11–50.

This informative paper was published after Efraim Fischbein passed away. The paper summarises his theory describing the evolution of intuitions, the relationship between intuitions and logical thinking, and discusses the educational implications of the theory.

Greer, B. (1987). Nonconservation of multiplication and division involving decimals. *Journal for Research in Mathematics Education*, 18, 37–45.

This interesting paper defines the phenomenon of nonconservation of the operations of multiplication and division. In it Greer clarifies that some pupils tend to choose different operations for juxtaposed word problems differing only in terms of the numbers involved. The lack of awareness of the invariance of the operations when only the numbers are changed should be taken into account when planning teaching (e.g. the 'substituting decimals by easy numbers' strategy is inefficient in this case).

References

Clement, J. (1993) Using bridging analogies and anchoring intuitions to deal with students' preconceptions in physics. *Journal of Research in Science Teaching*, 30, 1241–1257.

Fischbein, E. (1987) *Intuition in Mathematics and in Science: An Educational Approach.* Dordrecht, Netherlands: Reidel.

Fischbein E. (1993) The Interaction between the formal, the algorithmic and the intuitive components in a mathematical activity. In R. Biehler, R.W. Scholz, R. Sträßer and B. Winkelmann (eds), *Didactics of Mathematics as a Scientific Discipline* (pp. 231–245). Dordrecht, The Netherlands: Kluwer.

Fischbein, E., Deri, M., Nello, M.S. and Marino, M.S. (1985) The role of implicit models in solving verbal problems in multiplication and division. *Journal for Research in Mathematics Education*, 16, 3–17.

Flavell, J.H. (1977) *Cognitive Development*. Englewood Cliffs, NJ: Prentice Hall.

Fuson, K.C. (2003) Developing mathematical power in whole number operations. In K. Kilpatrick, G. Martin and D. Schifter (eds), *A Research companion to Principles and Standards for School Mathematics* (pp. 68–94). Reston, VA: National Council of Teachers of Mathematics.

Haylock, D. and Cockburn, A. (2003) *Understanding Mathematics in the Lower Primary Years*. London: Paul Chapman Publishing.

Kilpatrick, K., Martin, G. and Schifter, D. (eds) (2003) *A Research Companion to Principles and Standards for School Mathematics*. Reston, VA: National Council of Teachers of Mathematics.

Lester, F.K. (ed.) (2007) *Second Handbook of Research on Mathematics Teaching and Learning*. Charlotte, NC: Information Age Publishing.

Stavy, R. and Tirosh, D. (2000) *How Students (Mis-)understand Science and Mathematics: Intuitive Rules*. New York: Teachers College Press.

Tirosh, D. (2000) Enhancing prospective teachers' knowledge of children's conceptions: The case of division of fractions. *Journal for Research in Mathematics Education*, 22, 125–147.

Tsamir, P. (2003) From 'easy' to 'difficult' or vice versa: The case of infinite sets. *Focus on Learning Problems in Mathematics*, 25, 1–16.

Right or wrong? Exploring misconceptions in division

Pessia Tsamir, Sarah Hershkovitz and Dina Tirosh

> **Pause for thought**
>
> Are you a talker or more of a listener? Do you find that discussions help you clarify your thinking?

Over the years that we have worked first as school teachers and then as teacher educators, we have discovered how conversations with others are a very effective way to develop our understanding of a wide range of topics. Sometimes we come to such discussions thinking we are fairly knowledgeable. At other times we come with little or no understanding. Almost invariably, however, we come away from talking to friends and colleagues feeling better informed and energised.

In this chapter we will demonstrate how mathematical tasks can serve to stimulate discussions on mathematical conventions, mathematical operations and mathematical rules. That may sound rather dry but, we can assure you, lively conversations can often emanate from even the most formal-sounding topics. We have chosen division as our focus as it serves as both an end and a means. More specifically, as an end, division plays a prominent role in primary mathematics, and yet it is known to elicit many misconceptions. As a means, division is a good topic for exploring a range of issues surrounding children's difficulties and possible ways to address the challenges they face via 'follow-up questions' and 'next-step tasks'. Having said that, the strategies we propose can be used to explore a broad range of the mathematical topics covered in the primary curriculum. In essence, the technique we use involves asking children to solve mathematical tasks in more than one way, thus encouraging them to present and discuss different correct and incorrect ideas. We then challenge the class to investigate various thought-provoking solutions – correct and incorrect – some of which we will have prepared beforehand and others which may have been raised by the class.

Our discussion will focus on some segments of lessons that were taught by Tammy, a year 6 teacher, who frequently used her pupils' solutions as springboards for classroom conversations. In the lessons described here, Tammy wanted to highlight the characteristics of the operation of division, to promote her pupils' familiarity with mathematical conventions (such as the order of operations), and to improve their ability to use mathematical rules (such as the commutative and associative laws).

Our focus is on the double-division task $200 \div 50 \div 10$ that Tammy discussed with her pupils, after reading a related paper by Tsamir and Koren (2004). Before joining Tammy in her lesson, let's take a minute and think about the task she is about to present to her class.

Challenge

What is the solution of $200 \div 50 \div 10 = \square$?

Is there more than one solution?

Should the results necessarily be the same? Why?

Task 1: Solve in different ways

The class was asked to work individually and solve $200 \div 50 \div 10 = \square$ in at least two ways, and to explain their answers. Here is a segment of the lesson:

Danny: It's 0.4, because it's 4 divided by 10.
Yael: I got 40 ...
Ron: Yes ... so did I ...
T: Let's see who got what. Who got 0.4? [Several hands go up]
T: And who got 40?
Voices: Me ... me ... [again several hands go up]
T: Any other solution?
Gil: I got 4 ...
T: I see ... we have different suggestions ... different results ... perhaps there are more ways ...

In this short segment, the children came up with three different answers and thus it seems reasonable to assume that there were at least three ways in which the task was solved. Tammy then asked her class to work in small groups. The groups had to provide as many solutions as possible, to explain each answer in writing prior to submitting their work.

Pause for thought

In your opinion,

- what would be the children's common correct solutions to $200 \div 50 \div 10 = \square$?
- what would be the children's common incorrect solutions to $200 \div 50 \div 10 = \square$?
- what are the possible reasons for the children's incorrect solutions?

The pupils' solutions to $200 \div 50 \div 10 = \square$

Here are the solutions that the small groups came up with together with, the most typical explanations. (In case you are wondering, for the sake of clarity, the language used is ours not the children's!)

The correct solutions were as follows.

1. *Go from left to right:*

$$\underline{200 \div 50} \div 10 =$$
$$4 \quad \div 10 = 0.4$$

Explanation: When we have a chain of repeated operations and division appears more than once then the task should be done in steps, by working on each division at a time from left to right.

2. *Cancel two initial elements:*

$$200\!\!\!/ \div 50\!\!\!/ \div 10 = 4 \div 10 = 0.4$$
$$\underset{4}{\cup}$$

Explanation: When doing division with more than two numbers, we can reduce the initial two consecutive numbers by dividing them by the same number.

3. *Cancel first and last element:*

$$200 \div 50 \div 10 = 20\!\!\!/ \div 50 \div 1\!\!\!/0 = \frac{20}{50} = 0.4$$

Explanation: In division with more than two numbers, it is permissible to reduce the first and the last numbers by dividing them by the same number.

Turning to the incorrect solutions, as you might expect, these included a number of flawed ideas.

4. *Reduce everything:*

$$200 \div 50 \div 10 = 20\!\!\!/ \div 5\!\!\!/ \div 1\!\!\!/ = 20 \div 5 \div 1 = 4$$

Explanation: In division, as with fractions, we reduce the numbers by dividing *all* by the same number.

Here we see the erroneous *reduce everything* idea, which indicated that pupils did not differentiate the role of the first number (the dividend) from the other numbers (the divisors), thus for these pupils *all* numbers can be divided. They see it as permissible to cancel *the dividend* and *all of the divisors*.

5. *Reduce any pair:*

   $200 \div 50 \div 10 = 200 \div 5\not{0} \div 1\not{0} = 200 \div 5 \div 1 = 40$

   Explanation: In division, we can reduce any pair of consecutive numbers by dividing *both* by the same number.

In solution 5, we see the mistaken *reduce any pair* idea, which may indicate pupils' correct thoughts about fractions, where reduction may be done on a pair of numbers, i.e. on the nominator and denominator. As mentioned in the analysis of solution 4, here, one number should be the dividend and the other number should be one of the divisors.

6. *Compensate by adding a zero*:

   $200 \div 50 \div 10 = \underset{40}{\underbrace{20\not{0} \div 5\not{0}}} \div 10 = 40 \div 10 = 4$

   Explanation: It's okay to ignore the zero in 200 and in 50 temporarily so that we can solve $20 \div 5$. That equals 4. Then we have to put the zero back.

In solution 6, the pupils were probably mistakenly using a *model of multiplication*, or a model of decimals. In multiplication, for instance, 20×7 can be calculated in two steps: $2 \times 7 = 14$ and then add the zero to get $20 \times 7 = 140$, that is $20 \times 7 = 10 \times 2 \times 7 = 10 \times 14 = 140$. In the case of decimals, as in 2.3×4, it is possible to compute $23 \times 4 = 92$ and replace the decimal point to get 9.2, or in $13.2 \div 11$ you can calculate $132 \div 11 = 12$, and put back the decimal point to get 1.2.

7. *The 'associative' model:*

   $200 \div 50 \div 10 = 200 \div (50 \div 10) = 200 \div 5 = 40$

   Explanation: We did $50 \div 10$ first, according to the associative rule.

In solution 7, the pupils incorrectly applied the *associative law*. They overgeneralised from cases of double addition, $a + b + c = a + (b + c)$, for example $12 + 19 + 1 = 12 + (19 + 1) = 32$, and from cases of double multiplication, $a \times b \times c = a \times (b \times c)$, for example $12 \times 5 \times 2 = 12 \times (5 \times 2) = 120$, to 'double division', $a \div b \div c = a \div (b \div c)$, where it does not work: $200 \div 50 \div 10 = 0.4$, while $200 \div (50 \div 10) = 40$, thus $200 \div 50 \div 10 \neq 200 \div (50 \div 10)$.

8. *The 'distributive' model:*

   $200 \div 50 \div 10 = 10 \times (20 \div 5 \div 1) = 10 \times 4 = 40$

   Explanation: 10 is a common factor. So if I factorise the expression, with the distributive rule, I will have smaller numbers to work with in brackets.

In solution 8, the pupils incorrectly applied the *distributive law*. They overgeneralised from the distributing 'multiplication-over-addition'. Consider, for example, $10 \times (2 + 3 + 5) = 10 \times 2 + 10 \times 3 + 10 \times 5$, because $10 \times 10 = 20 + 30 + 50 = 100$. In other words, the strategy would work for multiplication but, in the example above, the children erroneously created a distributive law of 'multiplication-over-division', $m \times (a \div b \div c) = m \times a \div m \times b \div m \times c$, which does not work. In the above case, $200 \div 50 \div 10 = 0.4$, while $10 \times (20 \div 5 \div 1) = 40$, thus $200 \div 50 \div 10 \neq 10 \times (20 \div 5 \div 1)$.

Pause for thought

Would you present your pupils with various correct solutions? Why? Why not?

Would you present your pupils with incorrect solutions?

Why? Why not?

As you can see the children provided a wide range of solutions. Indeed Tammy was surprised by the sheer variety and unexpected richness of the information the children's work provided including, to her amazement, that the answers given by some of the groups included solutions resulting in different answers. She realised that, from the children's perspective, consistency, and the need to reach a single result to division tasks, was mistakenly not regarded as being necessary under the given circumstances.

Reflecting on this Tammy produced a second task for the class ...

Task 2: Right or Wrong?

In the following task (Figure 4.1), the children were asked to judge the correctness of ten solutions to $200 \div 50 \div 10 = ?$ as a preliminary step to stimulate class discussion.

Tammy decided to include two additional, correct solutions (8 and 10), which she devised herself and thought would be useful to discuss.

Challenge

What do you think Tammy hoped to gain by including these additional solutions?

The activity took place during a 90-minute class session, and consisted of three parts:

1. *Individual work.* The children were asked to decide which answers were correct and to write a detailed explanation to each solution.
2. *Group work.* In groups of three or four pupils each child discussed their opinions and dilemmas with a view to coming up with a 'group decision' regarding the

Figure 4.1 Right or wrong?

Right or Wrong?

For each solution, please say whether it is correct or not, and explain why.

Solution 1: 200 ÷ 50 ÷ 10 = □

4 ÷ 10 = 0.4

The solution is correct / incorrect (circle your choice). Why?

Solution 2: 200 ÷ 50 ÷ 10 = 200 ÷ 5Ø ÷ 1Ø = 200 ÷ 5 ÷ 1 = 40

The solution is correct / incorrect (circle your choice). Why?

Solution 3: 200 ÷ 50 ÷ 10 = 20Ø ÷ 5Ø ÷ 10 = 4 ÷ 10 = 0.4

The solution is correct / incorrect (circle your choice). Why?

Solution 4: 200 ÷ 50 ÷ 10 = 20Ø ÷ 50 ÷ 1Ø = 20/50 = 0.4

The solution is correct / incorrect (circle your choice). Why?

Solution 5: 200 ÷ 50 ÷ 10 = 20Ø ÷ 5Ø ÷ 1Ø = 20 ÷ 5 ÷ 1 = 4

The solution is correct / incorrect (circle your choice). Why?

Solution 6: 200 ÷ 50 ÷ 10 = 20Ø ÷ 5Ø ÷ 10 = 40 ÷ 10 = 4

∪

40

The solution is correct / incorrect (circle your choice). Why?

Solution 7: 200 ÷ 50 ÷ 10 = 200 ÷ (50 ÷ 10) = 200 ÷ 5 = 40

The solution is correct / incorrect (circle your choice). Why?

Solution 8: 200 ÷ 50 ÷ 10 = 200 ÷ (50 × 10) = 0.4

The solution is correct / incorrect (circle your choice). Why?

Solution 9: 200 ÷ 50 ÷ 10 = 10 × (20 ÷ 5 ÷ 1) = 10 × 4 = 40

The solution is correct / incorrect (circle your choice). Why?

Solution 10: 200 ÷ 50 ÷ 10 = ______ = 0.4

The solution is correct / incorrect (circle your choice). Why?

validity of each suggested solution, the reasons for its being correct or incorrect, and possible ways to determine validity. Each group had to prepare a written summary of their discussion, and to choose a representative to present the group's decisions and any questions they came up with to the class.

3. *Class discussion.* Tammy then conducted a discussion, highlighting correct and incorrect solutions, referring to reasons for the different judgments, and focusing on the need to understand the underlying notions (e.g. division and operation) and conventions (e.g. order of operating).

Tammy anticipated that having the children invest a considerable amount of thought on the different solutions, first individually and then in small groups, would be a good start to class discussion. She hoped it would promote children's awareness of their own knowledge of related issues, of their dilemmas, and of ways to communicate their opinions and their queries to the class.

Pause for thought

Would you adopt Tammy's strategy?

If you did, how do you think your pupils would respond?

In order to see of how her pupils' ideas changed during the series of activities, Tammy collected her pupils' individual sheets and the written summaries of the groups.

Reactions to the task

So how did Tammy's 11-year-olds respond to such a range of challenges? Looking through the work of both individuals and groups Tammy noted her pupils examined the validity of the ten solutions in three different ways:

1. *The 'result-test' approach.* The pupils either spotted a solution which they were certain was correct (usually solution 1 or solution 10), or solved $200 \div 50 \div 10$ for themselves and used their result as a test case. This was then compared with the results of the other solutions. So if, for example, a child decided that 0.4 was the correct solution all others with 0.4 as a result were automatically regarded as being correct and all solutions that had another result were automatically considered as wrong. In this way, consistency was preserved, because pupils decided in advance to preserve the result. However, the criterion of consistency is insufficient, the correct method is also required. For example, in another situation when pupils were asked to solve 2^2 one student wrote: 2×2, while another wrote $2 + 2$. Both got 4 as a result, but only the first solution is correct. Also, when children were asked to solve $2 \times (3 + 1.5)$ two children wrote $2 \times 3 \times 1.5$, while the others wrote $2 \times 3 + 2 \times 1.5$. Here too, the two solutions reach the same result, 9, but only the second one is correct.

Another drawback to this strategy was illustrated by one student who was convinced that 4 – rather than 0.4 – was the answer (solution 5) and thus accepted only incorrect solutions (i.e. those resulting in 4) while rejecting all the correct ones.

2. *The 'compartmentalisation' approach.* Here children treated each solution as a separate entity and made no attempt to compare the different results presented. Thus, they occasionally accepted different, inconsistent results. In other words, although these pupils paid attention to various correct and incorrect details in the different solutions, they missed the big picture of consistency, equivalency and uniqueness (singularity) of a result to a given operation.

3. *The 'same result is necessary but insufficient' approach.* These pupils used a combination of approaches 1 and 2. They had an initial solution that they trusted. However, on examining the other solutions, they were not satisfied with merely looking at the final results; they looked into the process of each solution, judging the requirement of reaching the same result as of paramount importance. Thus, solutions that seemed incorrect were rejected in spite of having the correct result.

This appears the most appropriate approach, and yet it is not error-proof. For example, several pupils followed solution 1, and all the related solutions except for solution 8. That is, they started by solving the task from left to right, continued by checking the process of each solution and also making sure that the result they reached was 0.4. However, they claimed that 'you cannot solve a division task by multiplication', or that 'division can only change to multiplication if you convert one number to a fraction, for example, $6 \div 3 = \square$ can be changed to $6 \times 1/3 = \square$', and therefore they rejected solution 8, although it had 0.4 as a result.

One child claimed that the correct solution is solution 5. In her view, this division task should be solved by reducing all numbers first. She further explained that solution 6 seemed incorrect, even though it had the same result, and that the other solutions did not have the same result, and therefore must be incorrect.

Pause for thought

What mathematical ideas do you think the children were not familiar with?

Are your pupils familiar with these ideas?

If you are unsure, how might you find out?

Tammy realised that if her class were to appreciate the dilemma of which solution is right and which is wrong, they needed to know that having three correct solutions – 0.4, 40, and 4 – was simply not possible. In order to address this, Tammy asked the children six questions:

(a) What is 'division'?
(b) What does it mean 'to reduce'?
(c) What are the conventions regarding the order of operations?
(d) What is the commutative law?
(e) What is the associative law?
(f) What is the distributive law?

In their discussion of these issues, the class concluded that division is a binary operation (i.e. it involves working with two numbers at any one time) on [real] numbers. They considered division in the following two ways:

> *Definition 1.* $a \div b$ $(b \neq 0)$ equals the *single number* c, so that $c \times b = a$.
> *Definition 2.* $a \div b$ $(b \neq 0)$ is the *single number we reach by* $a \times 1/b$.

The class then arrived at the following conclusion:

> When we get several results for a division expression, only one of the following two options is possible: *either* one of these results is correct, *or* all are incorrect.

On reaching this conclusion one of the pupils, Dan, said:

Dan: So, if we do ... a task ... division in more than one way ... and reach a certain result ... number ... the same number ... then we know that we got it right ...

T [to the class]: What do you think?

Voices: Yes ... I ... No ...

T: OK ... OK ... we have 'yes' and we have 'no' ... I need explanations ... you have to convince me ... yes or no?
[Several hands go up]

Anat: I changed my view. First, I believed that the 'reduce all numbers' was the correct way. So ... what was it ... solution number 5 OK ... there were two solutions, this and another one that came to 4: both were incorrect.

Gali: 40 appeared as an answer several times ...

T: So, when we get the same result several times, it might be that all solutions are correct, that some are correct, or that all are incorrect ... I see that we need to clarify what is 'to reduce'?

When discussing the idea of 'reducing', it was evident that the pupils believed that reduction was allowed in division. 'Reducing' was grasped as dividing the numbers by the same non-zero factor (number). Some pupils thought we should divide this number into all three of the numbers in the original task, some believed that it should be done on any pair of numbers in the task, while others believed that it should be done on specific pairs of numbers.

Hearing this, Tammy drew the children's attention to two types of numbers, the dividend and the divisor. In $200 \div 50 \div 10$ there is one dividend (200) and two divisors (50 and 10). She showed that if the dividend is divided by 2, it becomes $100 \div 50 \div 10$ giving half the original result. If we divide one of the divisors by 2 the result is twice the original one. So, dividing the dividend and one of the divisors by 2 (or by any number) preserves the result.

Pause for thought

How do you judge the solution

$$200 \div 50 \div 10 = 20\not{0} \div 5\not{0} \div 10 = 20 \div 5 \div 1 = 4?$$

What are your thoughts on the 'reduction in division' tasks?

In order to further these ideas, Tammy gave her class the assignment shown in Figure 4.2. The children's responses were as follows:

(1a) they realised that dividing the dividend by 5 gives a result 5 times smaller than the original one;
(1b) they observed that dividing the divisor by 5 produces a result 5 times larger; and
(1c) they grasped the idea that dividing both the dividend and the divisor by 5 maintains the result.

Figure 4.2 A division task with one divisor

Solve:

1) $200 \div 50 =$

(1a) $(200 \div 5) \div 50 =$
The dividend is divided by 5.
The result is ______________

(1b) $200 \div (50 \div 5) =$
The divisor is divided by 5.
The result is ______________

(1c) $(200 \div 5) \div (50 \div 5) =$
The dividend is divided by 5
and the divisor is divided by 5.
The result is ______________

Tammy then asked the children to complete the task shown in Figure 4.3. This involved situations where there is one dividend and two divisors.

Figure 4.3 A division task with several divisors

Solve:

2) $1600 \div 80 \div 20 =$

(2a) $(1600 \div 10) \div 80 \div 20 =$
The dividend is divided by 10.
The result is ______________

(2b) $1600 \div (80 \div 10) \div 20 =$
A divisor is divided by 10.
The result is ______________

(2c) $1600 \div 80 \div (20 \div 10) =$
A divisor is divided by 10.
The result is ______________

(2d) $1600 \div (80 \div 10) \div (20 \div 10) =$
Two divisors are divided by 10.
The result is ______________

(2e) $(1600 \div 10) \div 80 \div (20 \div 10) =$
The dividend is divided by 10
and a divisor is divided by 10.
The result is ______________

(2f) $(1600 \div 10) \div (80 \div 10) \div 20 =$
The dividend is divided by 10
and a divisor is divided by 10.
The result is ______________

(2g) $(1600 \div 10) \div (80 \div 10) \div (20 \div 10) =$
The dividend is divided by 10
and the two divisors are divided by 10.
The result is ______________

The children concluded that:

(2a) dividing only the dividend by 10 gives a result that is 10 times smaller than the original result;
(2b)–(2c) dividing one divisor by 10 gives a result that is 10 times larger than the original result;
(2d) dividing two divisors by 10 makes the result a 100 times larger;
(2e)–(2f) dividing both the dividend and one divisor by 10 keeps the result as is; and
(2g) dividing the dividend and two divisors by 10 makes the result 10 times larger.

One of the pupils, Edit, realised that she could see this connection more clearly by considering fractions:

> I can see it much better if I think about fractions ... I mean,
>
> $200 \div 50 \div 10 = 200 \times 1/50 \times 1/10$
>
> and it is clear that I am allowed to reduce pairs ... one top and one bottom every time.

The connection that Edit made between the two representations – the division of natural numbers and the multiplication of fractions – inspired Ron to use fractions in another situation:

Ron: Ahh ... now I can also understand the solution

$200 \div 50 \div 10 = 200 \div (50 \times 10)$

Because

$$200 \div 50 \div 10 = 200 \times \frac{1}{50} \times \frac{1}{10}$$

$$200 \div (50 \times 10) = 200 \times \frac{1}{50 \times 10} = 200 \times \frac{1}{50} \times \frac{1}{10}$$

T: Let's think about $200 \div 2 \div 5$ ... it's the same as $200 \div (2 \times 5)$, right?
Voices: Yes.
T: What's the answer?
Voices: 20.
T: There is a story I'd like to tell you. Once there was an old man, let's call him grandpa, who had two sons, and each son had five daughters. Grandpa had £200 which he wanted to share equally between his grandchildren. He thought: I have $2 \times 5 = 10$ grandchildren, so I have to divide my £ 200 into 10 parts. £200 ÷ (2 × 5) = £20. Then, he changed his mind and decided that he should give the money to his two sons. Each would receive: £200 ÷ 2 = £100. He would then ask each son to divide the money equally between his 5 daughters. That is, £200 ÷ 2 ÷ 5 = £20. If you stop to think about it, the choice of the numbers is not important as [writing on the board]

$A \div B \div C = A \div (B \times C)$ where $B \neq 0$, $C \neq 0$

Tammy ended the session by asking the children to summarise what they had learnt from the task $200 \div 50 \div 10$. The children's notes revealed the following:

- They learnt a lot from these lessons, and they had been very surprised because it was not that they studied 'new things' (their wording), but that they gained new understandings of familiar ideas such as 'division' and 'order of operations'. For example, they specifically mentioned the discussion of the need to have a single result to a division task.
- Several of them appreciated the discussion on cancelling factors, and others commented on their increased knowledge of the distributive and the commutative laws.
- Finally, all of the children said that they had appreciated the discussion of the equivalent expressions $A \div B \div C = A \div (B \times C)$ where $B \neq 0$, $C \neq 0$, and especially the everyday context that was added in the form of a story to show that 'it works'.

Summing up and looking ahead

Reflecting on the lessons, Tammy thought about ways to build on the ideas that had been discussed. Her plans are presented below, but first ...

Pause for thought

In your opinion, what were the key principles that Tammy used when designing her tasks?

Can you suggest similar, related, challenging tasks?

If you stop to think about it, in many ways subtraction 'behaves' similarly to division and so Tammy might decide to present the children with the task in Figure 4.4, which aims to show the following:

- If we have $a - b$, then $(a + 20) - b$ is 20 more than $a - b$ (e.g. $100 - 20 = 80$, but $(100 + 20) - 20 = 100$ which is 20 more than 80).
- If we have $a - b$, then $a - (b + 20)$ is 20 less than $a - b$ (e.g. $100 - 20 = 80$, but $100 - (20 + 20) = 60$ which is 20 less than 80).
- If we have $a - b$, then $(a + 20) - (b + 20) = a - b$ (e.g. $100 - 20 = 80$, and $(100 + 20) - (20 + 20) = 80$, thus $100 - 20 = (100 + 20) - (20 + 20)$).

In conclusion, adding the same number to both a and b leaves the result as is: $a - b = (a + c) - (b + c)$. Similarly, as with division,

1. $a - b - c \neq a - (b - c)$. For example, $100 - 50 - 20 = 30$, while $100 - (50 - 20) = 70$, thus $100 - 50 - 20 \neq 100 - (50 - 20)$.
2. $a - b - c = a - (b + c)$. For example, $100 - 50 - 20 = 100 - (50 + 20) = 30$.
3. $a - b - c = a - c - b$. For example, $100 - 50 - 20 = 100 - 20 - 50 = 30$.

Figure 4.4 Tammy's first subtraction task

Subtraction Task 1

a. Solve in two ways

125 – 67 – 37 =

b. Right or Wrong?

For each solution, please note whether it is correct, and explain why.

Solution 1: 125 – 67 – 37 = 125 – (67 – 37)
The solution is correct / incorrect (circle your choice)
Why?

Solution 2: 125 – 67 – 37 = 125 – (67 + 37)
The solution is correct / incorrect (circle your choice)
Why?

Solution 3: 125 – 67 – 37 = (125 + 3) – (67 + 3) – (37 + 3)
The solution is correct / incorrect (circle your choice)
Why?

Solution 4: 125 – 67 – 37 = (125 + 3) – (67 + 3) – 37
The solution is correct / incorrect (circle your choice)
Why?

Tammy then could give her class a second task (Figure 4.5) aimed to work on the following characteristics:

- If we have $a - b$, then $(a - 10) - b$ is 10 less than $a - b$ (e.g. 100 – 20 = 80, but (100 – 10) – 20 = 70 which is 10 less than 80).
- If we have $a - b$, then $a - (b - 10)$ is 10 more than $a - b$ (e.g. 100 – 20 = 80, but 100 – (20 – 10) = 90 which is 10 more than 80).

Figure 4.5 Tammy's second subtraction task

Subtraction Task 2

a. Solve in two ways

1225 – 640 – 225 =

b. Right or Wrong?

For each solution, please note whether it is correct, and explain why.

Solution 1: 1225 – 640 – 225 = 1225 – 225 –640
The solution is correct / incorrect (circle your choice)
Why?

Solution 2: 1225 – 640 – 225 = (1225 – 25) – 640 – (225 – 25)
The solution is correct / incorrect (circle your choice)
Why?

- If we have $a - b$, then $(a - 10) - (b - 10) = a - b$ (e.g. $100 - 20 = 80$, and $(100 - 10) - (20 - 10) = 80$, thus $100 - 20 = (100 - 10) - (20 - 10)$).

In conclusion, subtracting the same number from '*a*' and from '*b*' leaves the result as it is: $a - b = (a - c) - (b - c)$.

Figure 4.6 Tammy's final task

Operations Task

A. Solve each task in two (or more) ways

1. $125 \div 25 \div 5 =$
2. $240 \div 70 \div 7 =$
3. $1970000 \div 100 \div 100 =$
4. $1200 \div 120 \div 12 =$
5. $360 \div 200 \div 20 \div 120 =$
6. $240 \div 12 \div 360 \div 36 \div 100 \div 10 =$

B. Compare the following expressions

1. Is $3 \times 2 \times 1$ larger than / equal to / smaller than $3 + 2 + 1$?
   Why?
2. Is $30 \times 20 \times 10$ larger than /equal to /smaller than $30 + 20 + 10$?
   Why?
3. Is $360 \div 120 \div 20$ larger than /equal to /smaller than $36 \div 12 \div 2$?
   Why?

Tammy could produce a final task (Figure 4.6) which might draw the children's attention back to division and then go on to address different operations and highlight additional points related to the order of operations.

Challenge

Have a go at the tasks in Figure 4.6 and note any reactions you might have. How do they compare with the issues we discuss?

You may have discovered, the result B1 is quite surprising, because children often expect that the multiplication of numbers will have a larger result than the sum of these numbers, although this is not the case when one of the numbers is 1, a fraction or 0. For numbers larger than 1, the product is indeed larger than the sum, for example,

$$2 \times 3 > 2 + 3,\ 5 \times 4 > 5 + 4\ ,\ 80 \times 24 > 80 + 24.$$

But in the case of B1, $3 \times 2 \times 1 = 6$ and also $3 + 2 + 1 = 6$, thus $3 \times 2 \times 1 = 3 + 2 + 1$.

In other cases where one of the numbers is 1, a fraction or 0, we find, for example, that

$$2 \times 1 < 2 + 1,\ 10 \times 0.5 < 10 + 0.5,\ 3 \times 0 < 3 + 0.$$

You might expect several children to erroneously write that $3 \times 2 \times 1 > 3 + 2 + 1$. Interestingly, children might also write that $3 \times 2 \times 1 = 3 + 2 + 1$ (explaining that both are 6) but then continue and claim that $30 \times 20 \times 10 = 30 + 20 + 10$ (one pupil added, like in B1).

Question B3 was meant to examine whether the children had given up the erroneous *reduce everything* ideas.

In conclusion, this chapter illustrates an approach where children are asked, where possible, to solve mathematical tasks in more than one way. This provides a basis for rich thinking, and for the need to examine one's own solutions. Then we used Tammy's 'right or wrong' task as a basis for examining a class discussion of the different (correct and incorrect) solutions. We concluded with some examples of Tammy's possible next-step tasks designed to extend, consolidate and check the children's understanding.

Final challenges

A. What solutions do you predict your pupils would produce for the (i) subtraction tasks? and (ii) the operation task?
B. See if you can devise a 'right or wrong' task based on your pupils' solutions.

Further reading

Zaslavsky, O. (2005) Seizing the opportunity to create uncertainty in learning mathematics. *Educational Studies in Mathematics*, 60, 297–321.

This paper is a fascinating reflective account of the design and implementation of mathematical tasks that evoke uncertainty for the learner. Three types of uncertainty associated with mathematical tasks are discussed and illustrated: competing claims; unknown path or questionable conclusion; and non-readily verifiable outcomes. One task is presented in depth, pointing to the dynamic nature of task design, and the added value stimulated by the uncertainty component entailed in the task in terms of mathematical and pedagogical musing.

Reference

Tsamir, P. and Koren, M. (2004) The story of a task. *Mispar Hazak*, 7, 39–43 [in Hebrew].

5

Developing an understanding of children's acquisition of number concepts

Anne D. Cockburn

Sometimes teaching mathematics can be wonderfully exciting and stimulating. You pose a problem and the children run with it and come up with all manner of interesting strategies and solutions: the classroom becomes alive with excited exchanges and possibilities. Sometimes – even for the most expert mathematicians – teaching mathematics can be a challenging and somewhat mystifying experience. You plan a lesson in detail, carefully ensuring step-by-step progression, and seemingly bizarre answers and baffling conversations result. In this chapter I wish to explore why this might come about and how we can enhance the experience of teaching mathematics in primary school classrooms. Using topics introduced in earlier chapters, I will refocus our attention more towards issues of communication rather than the underlying mathematical content.

Issues of language and how misconceptions may arise through misunderstandings

One of the most commonly cited reasons why children find place value difficult is the language we use for the numbers between 10 and 20. Linda (year 2) observed that:

> There is not enough time spent on 10 to 19. Teen numbers are the sticky wickets since they don't fit the pattern of the language. When chanting/counting back the teens are often missed out completely. For example when going back in 5s from 30, 25, 20, 10, 5, 0.

More specifically, the early numbers – 11, 12 and 13 – tend to be unfamiliar and therefore offer no clues as to how you might say or write them. It is the numbers 14 to 19 which offer the most difficulty, however, because their names are said back to front: saying 'fourteen', for example, may lead one to think in terms of '4' followed by the conventional representation for 'teen' (i.e., '1'). Not surprisingly, many children confuse numbers such as sixty-one and sixteen as they both start with 'six'. Life might be easier if we taught in Welsh rather than English as there are regular, predictable patterns in the way one says the numbers, with 11 being *un deg un* (literally one ten one), 12 being *un deg dau* (one ten two) and so on. This will be discussed further below when we consider oral and written work, but the important point here is that, as far as possible, you ensure that you and your pupils are sharing a common understanding. This is often not as easy as it sounds, for a child's response to a question may provide insufficient – or even inaccurate – information. The task for the teacher is to discover which, so that the quickest and most effective action might be taken. When Ivan asked Ed in his year 1 class whether the digit 2 is worth more in 12 than in 20, Ed replied, '12 as it is worth more than 20. It is a bigger number.' The conversation, however, did not end there:

Teacher: Would you rather have 12 or 20 sweets?
Ed: 20.
Teacher: Why?
Ed: Because it is a bigger number.

When Ivan later asked a group of children, 'What is one less than ten?' he was given a range of responses from 1 to 21. Being rather surprised by the children's answers he asked a follow-up question, 'What does one less mean?' The replies to both questions are given in Table 5.1.

Table 5.1 Year 1 children's responses to 'What is one less than ten?'

Child	What is one less than ten?	What does one less mean?
Eva	'9'	'It means take one away'
Rob	'1. One is the first number and it is less than ten'	'Less means take away'
Joe	'21' (but wrote '12')	'One more added'
Ed	'8. It is less than ten'	'A number far away'
Alice	'9 because it is the number before 10. If you had 10 sweets and you ate one you would have 9.'	'It means take one away.'

Only one of the children, Alice, correctly responded to 'One less than 100?', but when the teacher rephrased the question as 'What number comes before 100?' three more children gave the correct answer. On reading this our Italian colleague, Carlo, wrote:

> This example is a 'masterpiece' for the conflict between the cardinal and ordinal approach. The ambiguity of 'less than' completes the picture of the situation. Less than, when re-expressed as 'the number before', is ordinal, less than as one minus is the cardinal approach.

In a busy classroom environment the temptation is to only stop to think about issues such as children's interpretations and mathematical meanings when one is faced with a surprising – almost invariably incorrect – response from a child or looks of incomprehension. To be honest, that is all most of us have time for as, coupled with our teaching of mathematics, we deal with all manner of other matters such as timing, behaviour and resource management. Popping an unexpected and challenging question into the conversation from time to time – even when the class are coming up with the correct responses – can be both revealing and make sessions more stimulating, informative and effective for the majority. Reflecting on earlier chapters, for example, we noted several cases where the 'right' answer was presented but faulty logic (which would later produce difficulties) was used to produce it. The use of appropriate language in such exchanges, although not always easy to generate spontaneously, helps considerably both for ease of communication and for building future conceptual understanding. Inexperienced teachers, in particular, often find it helpful to plan the precise mathematical terms they will use prior to a session in order to ensure that (a) *they* know what they are talking about and (b) they check that the children share their understanding of the terminology used.

Ruth (a year 2 teacher) explains that, as a result of our project, she

> will now be extremely careful re vocabulary and using a range of vocabulary. I suppose just being more aware of possibility that of a class of 30 children there may be as many of 7 or 8 who do not understand.

Teachers Ruth and Ivan now phrase questions in a range of ways and check on their pupils' understanding – even when they answer correctly – by asking several questions on the same topic. Quite unexpectedly they stumbled across a further factor which may influence a pupil's response.

Pause for thought

Compare and contrast the answers given in Tables 5.2 and 5.3 and see if you reach the same conclusion as Ivan and Ruth.

Looking at the responses in Tables 5.2 and 5.3, you may or may not be surprised that the children's initial responses were not 'yes' or 'no'. Interestingly, that never occurred to us until a colleague pointed it out. If you were asked a question such as 'Is there more fat in full-fat milk or semi-skimmed?', your answer is rather more likely to be 'Yes' than 'Full fat'. Reflecting on this further strengthened our belief that children often become familiar with the type of questions we ask and respond accordingly. Ivan, for example, typically asks either/or questions, giving pupils two options and expecting one or other in response. It may be that, as a result, Bob, Luke and Ross felt that only one number could be given in response to 'Is the digit 9 worth more in 19 than 9?'. Interestingly, however, as can be seen in Table 5.4, Bob responded differently when asked, 'Is the digit 9 worth more in 19 than 89? Why?'

Table 5.2 Year 1 responses to 'Is the digit 9 worth more in 19 than 9?'

Child	Is the digit 9 worth more in 19 than 9?	Why?
Bob	19	Because 19 has a ten in it
Luke	19	Because 9 hasn't got a ten but 19 has
Ross	19	Because it is a bigger number

Table 5.3 Year 1 responses to, 'Is the digit 2 worth more in 28 than 92?'

Child	Is the digit 2 worth more in 28 than 92?	Why?
Bob	28	Because in 28 it means 2 lots of 10
Luke	28	Because 28 has more units but 92 has more tens
Ross	28	Because it is in the ones column for 92 and the tens column for 28

Table 5.4 Year 1 responses to, 'Is the digit 9 worth more in 19 than 89?'

Child	Is the digit 9 worth more in 19 than 89?	Why?
Bob	Both the same	Because of the nines being in the units column
Luke	89	Because 89 has more tens
Ross	89	Because 89 is a bigger number

We do not know exactly why Bob answered one problem incorrectly and then seemingly saw the light and succeeded with the next question. It may be, however, that he stopped to think about what was really being asked of him rather than quickly responding in a rather automatic manner. As teachers it is sometimes hard to ascertain our pupils' thinking, but it is worth remembering that, although both you and your class might both be 'on task', you might in fact be focusing on different angles. For example, moving away from place value for a moment, when explaining that, mathematically, $3 \times 5 = 5 \times 3$ (i.e. that multiplication is commutative), a teacher might ask children to imagine two buildings: one with 3 floors and 5 rooms on each and another with 5 floors and 3 rooms on each. Both teacher and pupils might reach the conclusion that both buildings have the same number of rooms and one might conclude, from a mathematical perspective, that this implies that $3 \times 5 = 5 \times 3$. In contrast some, or even all, of the children might reach the very opposite conclusion, imagining the buildings the teacher described rather than the

Figure 5.1 Two buildings with the same number of rooms, illustrating the commutativity of multiplication

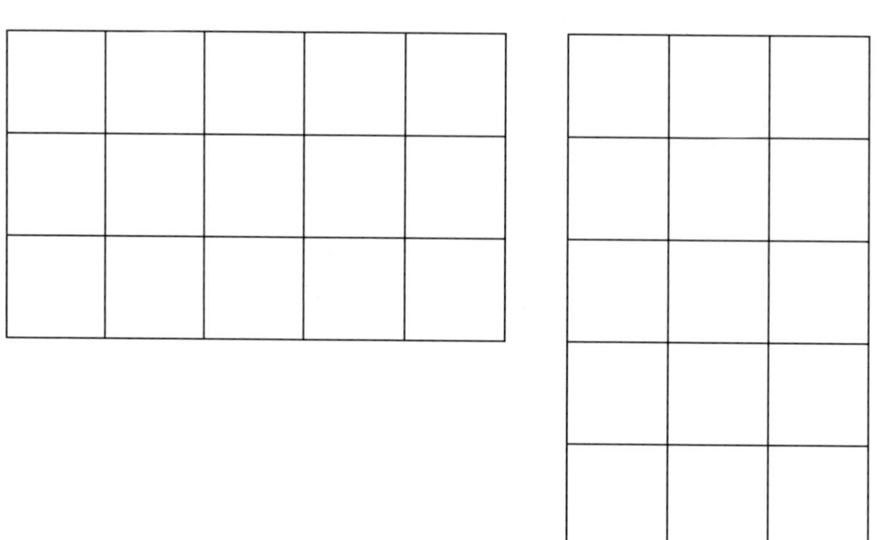

mathematics: the left-hand building in Figure 5.1 looks very different from the right-hand one.

The situation could become further complicated if the more creative children in the class started picturing doors and windows in the buildings, making it even more difficult for them to suspend their imaginations in order to, for example, 'turn the building on its side'. As primary teachers of English we actively want to encourage children to use their imaginations, which makes finding appropriate analogies quite a tricky operation. Starting with a small wood – such as illustrated in Figure 5.2 – seems less fraught with difficulties than flats when it comes to illustrating commutativity.

Figure 5.2 A wood illustrating the commutativity of multiplication

In brief, as you may have noted reading this book, we often use everyday examples to help illustrate mathematical ideas in more accessible ways. This will be discussed further below, but the important point here is that if you use analogies, you ensure that you and your pupils adopt the same perspective and that, as far as you can tell, they are mathematically sound.

Language, writing and reading and further scope for misunderstandings

Talking about numbers may be confusing, but the situation becomes trickier when committing them to paper. Linda explained a common problem:

> 21 and 12 is a classic. Even if they are written side by side some year 2s will still read both as 'twelve'.

Strangely, however, it does not seem to affect the way most of her pupils operate – if she points to the numbers and asks which the children would prefer if they were quantities of sweets, the majority point to 21 rather than 12.

Similar problems can arise in year 3, with some children demonstrating a facility with numbers but not being able to write them. For example, Jack knew that ninety-nine add two was one hundred and one but wrote as his response '1001'. As discussed above, therefore, it is important to ask several questions – preferably combining oral and written responses – to ascertain pupils' understanding.

Difficulties in reading and writing large numbers seem to continue throughout the primary years for some children. Kath notices that even some of her highest attainers in year 6 find writing numbers such as ten thousand and seventeen hard, frequently recording it as 1000017. She thinks that this is due to:

> lack of practice/experience and general underlying problems with place value and not being sure of the value of the columns after you pass thousands.

Challenge

Without pausing to think, how would you write the number 'three hundred and ten million, six thousand and ninety-five' and how would you say '4230967805'?

In both cases, were you able to provide instantaneous responses? If not, how did you work them out? In my case, with 'three hundred and ten million, six thousand and ninety-five', I started with the first (i.e. the left-hand) digit, '3', and then gradually worked towards the right-hand side of the number, with regular adjustments to the left as I went.

In contrast, with '4230967805', I started by grouping the digits in threes from the right-hand side which resulted first in '805' followed by '967 805' and ended in '4 230 967 805'. It was only then that I could contemplate reading the number aloud. Neither of the above tasks proved particularly easy for me and yet I consider myself to be reasonably proficient at place value!

Having said that, we are relatively lucky: just think how much more complex it is for Israeli children who read words from right to left but numbers from left to right. We are also lucky in that we have relatively few number names to learn to say, read and write; the French, on the other hand, have to master numbers such as *quatre-vingt-douze*, which means four-twenty-twelve or 92.

Getting a feel for number

Many of the examples described are of children who have not yet developed a feel for numbers. Sperry Smith (2001: 110) defines number sense as

> using common sense based on the way numbers and tools of measurement work within a given culture. It involves an appreciation for the reasonableness of an answer and the level of accuracy needed to solve a particular problem. It helps students detect errors and choose the most logical way to approach a math challenge.

What can we do to enhance this? As discussed in the introduction to this book, we know that 3- and 4-year-olds can have had a range of experiences with number and, in the teacher's view, have a sound understanding of the basic principles. Some of those described, for example, could recognise the numerals up to 20 and most could count with their teacher up to 26 (i.e. class size).

Pause for thought

How often do you ask your class to count quantities beyond 20 items? Have you ever shown them more than 1000 objects and discussed the quantity involved?

I am not proposing that children spend hours and hours counting out large quantities, but simply that they come to appreciate what a large number of, for example, one-penny pieces looks like. Depending on the stage children are at, you could ask each child in the class to count out 100 and then gradually put the piles together, discussing the size of the growing pile. Watching how the children amass their specified quantity could also give you an indication of their feel for number. For example, someone who is comfortable with number may count out a row of ten items, add a single column of ten and then arrange items in columns as shown in Figure 5.3 until the matrix is full. In other words, they can visualise what is required and use this knowledge to find the most efficient way to complete the task successfully.

Ian (year 1) is very keen on such activities, explaining that young children do not see large quantities and therefore when they set eyes on, for example, 100 beans for the first time, it is hardly surprising that they record them as '10,000'. Given plenty of practical experience, however, they begin to appreciate that

Figure 5.3 A child's system to assist in the counting out 100 one penny coins

0	0	0	0	0	0	0	0	0	0
0	0	0	0						
0									
0									
0									
0									
0									
0									
0									
0									

100 items, although a large quantity, is very different from 1000, let alone 10,000. We observed that such a real lack of understanding of the concept of large numbers extends well beyond the early years of schooling. Kath (year 6) reported that her pupils made 'wildly inaccurate estimates of populations of local villages/towns/cities and distances'. She went on to say that she felt that the children had 'no feel for the numbers' and described how, on being asked to describe a million, one child wrote 'Like the population of Amrica [*sic*] and the United Kingdom.'

A child's feel – or lack of feel – for numbers can often be detected using number line tasks. When asked to place specific numbers on a partial number line, for example, a low-attaining year 4 child, Lucy, responded thus:

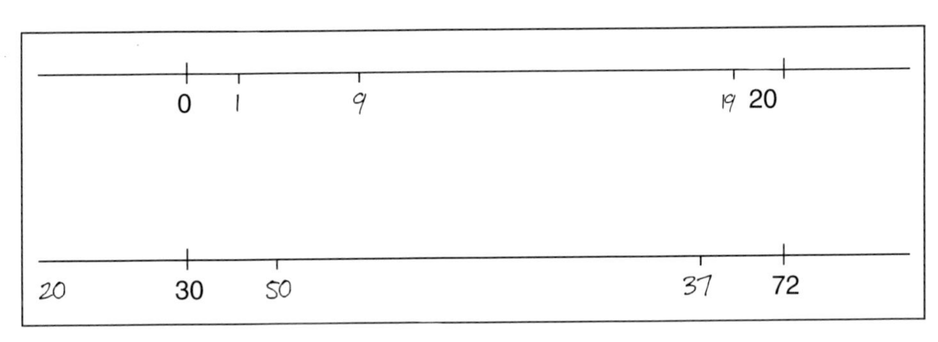

From these her teacher deduced that Lucy was far from secure with numbers beyond 20, but that she may well have a promising foundation on which to build.

As an aside, we noted that children's responses to the tasks we set often reflected their teachers' teaching strategies. For example, both Ruth and Linda's classes of 6-year-olds were asked to place '9' and several other numbers on a 0–20 number line as described above. In discussing it later, Ruth concluded:

> Most children knew that 9 came about the middle although some said 'just over' the middle and others said 'just under'. Only one child was out of his depth, saying 'quite a bit under the middle'.

In contrast, Linda observed:

> Most children found 9 the hardest, placing it too close to 0 or too close to 19.

One of our students had children completing a similar task using string number lines stretched between two chairs in the classroom. Zero and 100 were fixed, and children were required to place numbered clothes pegs (representing 15, 35 and 65) at appropriate points. Starting at '0', Pauline wrapped her left hand around the string then placed her right hand curled up beside it. Her left hand then leap-frogged her right and the peg marked '15' was placed directly to the right of it. A similar process was adopted for placing '35' and '65'. As it happens, the string was not 20-hands worth long but, nevertheless, Pauline's logic suggests a good understanding of how numbers relate to one another on a number line.

Examples our colleagues Graham and Darina reported show that, although upper primary children might appear to have a good understanding of whole numbers, this is not always apparent when decimal numbers are introduced. For example, they noted that 6.25 – 4 = 6.21 was a fairly common response and that children came up with one of three responses – 4.04, 4.4 and 4.94 – when faced with 5.07 – 1.3 = □. As will be discussed later in the book, these responses support the case for further work on the number line, including relating it to plenty of practical examples (see below) involving, for example, length and capacity.

The problems with patterns

As we note several times throughout this book, patterns are highly important in mathematics. How to develop a sound understanding of appropriate patterns will be discussed in Chapter 6, but here we wish to illustrate some of the unfortunate generalisations we have observed to demonstrate just how easily they occur and how we might avoid them.

When toddlers start learning to speak they tend to generalise their knowledge to build new words. Thus, for example, on appreciating that 'walk' and 'laugh' become 'walked' and 'laughed' when using the past tense, young children might use this knowledge to create 'seed' from 'see' and 'goed' from 'go'. Five-year-old Sandy illustrated an interesting variation on this. Her task was to count out 50 bricks. She managed 1 to 24 with no difficulty, and then continued 'Fifty ten, fifty eleven, fifty twelve' at which point her teacher intervened. On hearing this, Kath (year 6) explained that:

> A common problem is that whilst pupils are good at identifying patterns, they tend to apply them to other situations without considering their appropriateness.

She then described a common misconception among her pupils: when counting in decimals they often recite '1.6, 1.7, 1.8, 1.9, 1.10, 1.11 ...'. She addresses this situation using number sticks and number lines.

The fact that some children apply their knowledge of pattern – albeit possibly incorrectly – is to be commended, but sometimes, I would suggest, adults exacerbate the situation. In this particular situation, it may well be that the children had simply become used to thinking about money in terms of so many pounds and so many pence, recording the information incompletely (e.g. £1.6 instead of £1.60) or inaccurately (e.g. £1.6 rather than £1.06) and then overgeneralising this approach when working with decimals. More generally, we suspect children may think in terms of inappropriate patterns because:

- we expect too much. Laura (year 3), for example, suggests that:

  > We spend so long with two-digit numbers then suddenly expect them to be able to apply all that they know to thousands and tens/hundreds of thousands.

  Indeed, I suspect, we are often very thorough when it comes to explaining and working with numbers up to 20 and then – albeit unconsciously – we rather assume that the numbers 21–99 will take care of themselves. Another case of expecting too much is seen in the case of James who could add single-digit numbers correctly but had insufficient understanding of how to add three-digit numbers, let alone decimals, with the result that he concluded that 2.64 + 3.74 = 5.138.

- we give children tips such as 'when you multiply by 10 you just add a zero', 'when you subtract you always begin with the bigger number', 'when you divide you always end up with a smaller number than the one you started with' and so on. The temptation to introduce such 'rules' is particularly high for some parents and Kate (year 6) finds it very frustrating especially when 'we try so hard to be consistent at school and to always teach with understanding'. One of the most effective strategies for dealing with such situations is to introduce practical examples which blatantly do not follow the 'rule' which has been taught. For example, if you divide four cakes in half, how many pieces of cake will you have?

Challenges

Think of examples (1) to illustrate that adding a zero when multiplying by 10 does not always work and (2) to demonstrate that you do not always start with the bigger number when you subtract.

Interestingly, both children and parents often enjoy, and respond well to, such counter-examples as it gives them new insight into mathematical processes. The art for teachers, of course, is to pose such examples as motivating challenges rather than opportunities to demonstrate ignorance.

The fact that some children will overgeneralise their learning is inevitable and, as we want to encourage them to look for patterns in mathematics, it should not be of undue concern. Having said that, we suggest that not only are your pupils discouraged from applying quick tips to solve mathematical problems, but also they are encouraged to check their work using alternative routes.

The latter includes mathematical strategies such as completing a problem in one of two ways. For example, I could check whether $0.5 \times 10 = 5$ or 50 or, even, 0.5 by completing $0.5 + 0.5 + 0.5 + 0.5 + 0.5 + 0.5 + 0.5 + 0.5 + 0.5 + 0.5 = \square$. As discussed below, another strategy to determine whether a response is correct or not is to convert it into an everyday example.

Finally, the likelihood of inappropriate generalisations occurring is much reduced if children have very firm foundations in the principles of number. This is discussed further in the final chapter.

The complexity of context

Putting mathematical questions into everyday contexts is usually helpful and often revealing. For example, 'Put these numbers in order, with the least first: +2, −1, −4, −8, +6' is much easier if considered in terms of temperatures – 'Put these temperatures in order, with the lowest first: 2°C, −8°C, −1°C, 6°C, −4°C'. Similarly, fractions can more easily be visualised if presented in terms of pieces of pizza bought for a group of friends rather than abstract calculations such as $\frac{1}{4} \times 10 = \square$.

Care must be taken, however, to consider when specific everyday contexts might best be introduced. Money is a case in point. We have found that young children are often confused when asked to complete exercises involving exchanging 10 one-pence pieces for a ten-pence piece. If, however, they have already had plenty of experience of exchanging 10 unit blocks (as in the case of Dienes blocks) for a ten rod as shown in Figure 5.4, they are far less likely to be confused when this understanding is extended to the more abstract representation of money.

Figure 5.4 Exchanging ten ones for a ten rod

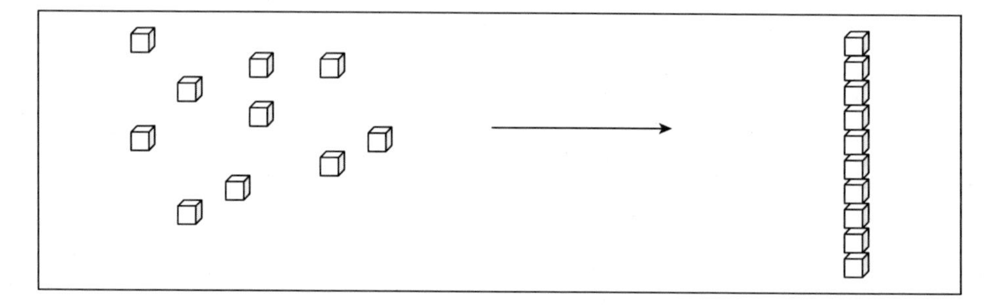

The way in which we discuss money can also create confusion for children, as teachers Donna (year 4) and Linda (year 3) discuss:

Linda: Decimals are introduced in the context of money but this happens before we even begin to approach how this is a natural extension of place value.

Donna: We don't make the links between money and the place-value system.

Linda: I think children find it confusing because the tenths are tens and the hundredths are called pence.

Donna: I agree, and this creates problems when children record amounts of money so we commonly get £1.5 instead of one pound and five pence.

Linda: It doesn't help that in mathematics we say one point five seven but in money we usually describe the same figures as one pound fifty-seven.

Talking about mathematics in general, Laura (year 2) observed that:

> Children are very literal so we need to be so careful with the language we use. You need to be very specific.

When confusing examples of language arise, as when using money or time for example, it is often helpful to draw children's attention to the conventions used so that they can begin to understand what might otherwise appear to be inconsistencies. This, in our view, however, should only be done when children have a sound knowledge of place value to which they can refer. Thus, for example, they would know that 1.5 was equivalent to 1½ and could be represented as 1½ pies and that, therefore, £1.50 must represent £1½ rather than one pound and five pence.

Challenge

What aspects of time might be confusing for children with an insecure knowledge of place value? Why?

Seeing the broader picture

As described at the beginning of this chapter, young children often appear to have quite an extensive knowledge of the numbers 1 to 10: they can generally recite them and recognise them, and they may even be able to write them. Before they can move on to understanding place value successfully, however, it is important that they can make connections between these various activities. Haylock (Haylock and Cockburn, 2008) expresses this need using the model presented in Figure 5.5. Many students and teachers have found this to be an extremely helpful way in which to consider the development of sound mathematical conceptual understanding. Reflecting on the above discussion, it is important to emphasise that the younger the child, the closer the examples need to be to the child's experience and interests. Returning to the honey collecting described earlier in the book (Cameo 2 in the Introduction), for example, consider the following. When

Figure 5.5 Making connections

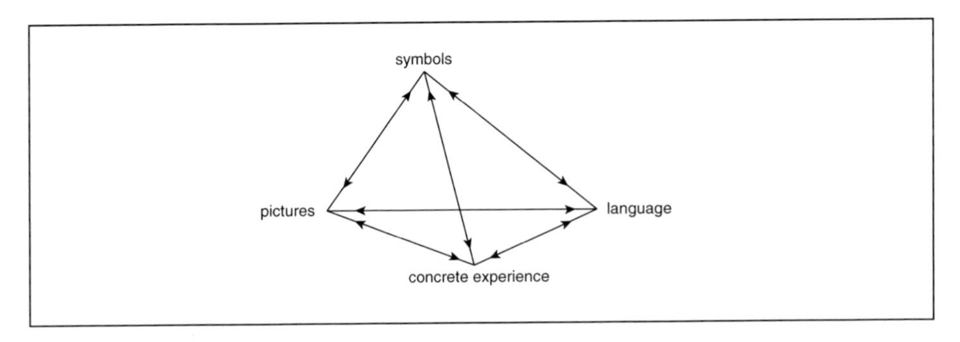

it was Colin's turn to collect some pollen, he seemed unable to recognise '4' on the card the teacher gave him and the following conversation took place:

Teacher:	Let's count them together [turning the card over to reveal 4 dots]
Teacher and Colin:	1, 2, 3, 4
Teacher:	How old are you?
Colin:	[holding up 4 fingers] 4

Colin then went off and stuffed his bucket full of bricks. On his return the teacher showed him the card he had in his hand (i.e. '4') and asked him what number it had on it. Colin replied '5'. The teacher then asked:

Teacher: Can you remember how old you are, Colin?
Colin: 4.
Teacher: Can you put out four bits of pollen (bricks)?

Colin put a brick on each of the dots on the card. He and the teacher then decided that he should return the other 'bits of pollen' to the flower where he found them.

Associating Colin's age with the numeral '4' proved a powerful reference point for him and helped reinforce the multiple representations of 'four' under consideration. The introduction of a personalised reference seemed to be the clue to helping Colin develop a broader understanding. Such an approach is implicit in Haylock's model, but – for beginning teachers in particular – I would suggest that explicit reference to the child's world is crucial. As children become older such highly specific references usually become less necessary, for they have a broader base of experience on which to build. I would argue, however, that, regardless of our age, if we are struggling with a new concept, the closer our teacher can come to describing it in terms with which we are personally familiar the better. Thus, for example, when learning a new language, I find it much easier to pick up words which have a familiar ring to them than those that do not. The Czech language is a case in point: after several visits to Prague I have at last mastered *ahoj!*, which is pronounced 'ahoy' and means 'bye'. Sadly, most other words seem beyond me as they sound so very different from anything I have heard before.

Related to this idea is to recognise that different people tend to learn in different ways and, indeed, that we learn in different ways depending on the context. Some people respond better to pictures than to language, for example. I usually prefer to learn practical things – such as driving – by doing them myself, but sometimes I learn more if I observe someone else in action – on the computer, for example. When introducing learners to a challenging concept for the first time, therefore, it is a good idea to reflect on the best mode of initial presentation and then extend the ideas – much as the teacher did above – to incorporate other aspects of Haylock's model.

Concluding remarks

In this chapter I have explored some of the complexities of teaching mathematics in primary classrooms and how misconceptions can arise for a variety of reasons. Through the use of judicious questioning and observation, however, I propose that a shared understanding between teacher and pupil is more likely to develop. This, together with the use of thoughtfully chosen child-centred examples, can help reduce the likelihood of children making false deductions and arriving at inappropriate mathematical understandings. Having said that, caution needs to be exercised, for even the most experienced teachers can be caught unawares when they ask a non-standard question. Much to the amazement of his teacher, a year 6 child recently explained 'One less than 210 is 110' – clearly he had not mastered place value quite as well as previously thought.

Summary of key ideas

- Children may respond to questions without a full understanding of what is being asked: check that you share a common vocabulary.
- Endeavour to ask a variety of different types of questions to encourage children to really reflect on what is being asked of them.
- Some situations can be viewed from more than one perspective: check that you and your pupils adopt the same perspective.
- Seeing patterns is an important skill in understanding mathematics: encourage your pupils to search for patterns but ensure they spot mathematically sound examples.
- Relate new ideas and concepts to what the children already know and understand.
- Provide children with plenty of practical experience of mathematics in meaningful contexts.

Anghileri, J. (2000) *Teaching Number Sense*. London: Continuum.

This is a very readable book which specifically focuses on the connections children need to make in developing number sense. It includes many practical examples suitable for teachers across the primary age range.

Dehaene, S. (1997) *The Number Sense: How the Mind Creates Mathematics*. London: Allen Lane.

Stanislas Dehaene covers a wide range of interesting topics in his book, such as whether numbers have colours, the limits of animal mathematics, and the adult number line. You might not wish to read it all, but the book is full of accessible and useful information, including why Piaget's findings on conservation of number are not necessarily as reliable as you might originally have thought.

Nunes, T. and Bryant, P. (1997) *Learning and Teaching Mathematics: An International Perspective*. Hove: Psychology Press.

Some might consider this to be a fairly academic book, but it includes some very readable chapters which encourage you to consider children's perspectives of mathematics.

Ginsburg *et al.*'s chapter, 'Happy birthday to you', provokes some important questions for those working in the multicultural classroom, while Streefland's work on fractions, 'Charming fractions or fractions being charmed', explains why real-life, child-centred examples are so helpful in helping children appreciate the mathematics of fractions.

Turner, S. and McCullouch, J. (2004) *Making Connections in Primary Mathematics*. London: David Fulton.

As the title implies, this book discusses the importance of making connections to enhance mathematical understanding. It is full of practical suggestions and has some good ideas on incorporating mathematics in cross-curricular work.

References

Haylock, D. and Cockburn, A.D. (2008) *Understanding Mathematics for young children* (3rd edn). London: Sage.

Sperry Smith, S. (2001) *Early Childhood Mathematics* (2nd edn). Boston: Allyn & Bacon.

Highlighting the learning processes

Graham Littler and Darina Jirotková

Pause for thought

Do you ever find yourself wishing you had more time to discuss a mathematical topic with your pupils? If so, what do you do in such cases? If not, how do you manage to ensure everyone has a good understanding of the work before you move on to the next concept?

From time to time most of us find that the pressure to get through the curriculum forces us to work at a greater pace than we would like. Thus, for example, we sometimes find we have to move on to the next topic before everyone in the class has had time to fully grasp the concepts involved in the current one. As a result we may resort to shortcuts and handy tips so that those who have not understood all the concepts we have been working on can at least develop algorithms, procedures which can be memorised to solve particular problems, such as 'multiply by 10, add a zero'.

Often learners are delighted to be given such algorithms as it usually means quick and – at the time – almost inevitable success. Such solutions, however, give a false sense of security and understanding. It is only later when they encounter more advanced, or non-standard, mathematics that children begin to appreciate that these quick-fix techniques do not always work and, indeed, that they have little or no understanding of the mathematical concepts involved.

Challenge for teachers

Thinking back to your own school days, can you remember times when you were given mathematical tips and hints (mechanical knowledge)? Did they help or hinder you at the time? Do you still use some of them? Why do they work? Do they always work?

Through the presentation of practical examples we will introduce a theory developed by Milan Hejný (Hejný and Kratochvílová, 2005) to help teachers to design activities so that their pupils develop a sound understanding of the learning process of mathematical concepts. Using cases which we observed across the four project countries, we will then illustrate how Hejný's theory helps teachers to develop their teaching strategies to enable children to gain a deeper mathematical understanding and, in so doing, lessen the chance of them developing misconceptions.

Illustration 1: *Multipying by 10*. Picture the case of a teacher, June, who has been working with her class on the concept of multiplying by 10. At the beginning of the week when the topic is first introduced, there seems to be plenty of time for discussion and practical examples. As the days go by, however, June becomes increasingly aware that she is running out of time and that, if she is going to maintain her schedule, she must move on to the next topic despite the fact that not everyone has grasped the underlying concept of multiplication by 10. Rather than abandon the topic abruptly she therefore says to the class, 'I am sure some of you have noticed that when we multiply by 10 a zero is added to the right-hand side of the number we are multiplying. So a simple rule to follow is: "when we multiply by 10 add a zero!"'

Typically in such situations we have observed that those pupils who have not followed and understood the development of the concept will grasp a mantra as though it were a life jacket and treat it as a rule. To their relief they will find that it works for such exercises as 17 × 10. They then discover that they are less successful when suddenly faced with the problem 'I bought 10 pens and each cost £1.15, how much did I have to pay altogether?'. By using the rule they write £1.15 × 10 = £1.150, only to be marked incorrect. Thus, by applying their limited knowledge mechanically, pupils demonstrate that they do not understand the basic idea of multiplying by 10 and place value: they followed the rule slavishly without realising that it only can be applied in very restricted circumstances.

Illustration 2: Area of rectangle. Now let us move on to another scenario we recently observed. In this example three 9-year-olds – Sally, Abigail and Jane – gained knowledge by the realisation that all the experiments they had been working on had a common process. Their teacher had given them several tasks which involved finding the areas of various rectangles using non-standard units, squared paper, square tiles and so on by actually counting the units of area they were using. When we first encountered them they were finding out how many rectangular blackboard cleaners covered the blackboard. They observed that if they banged the cleaner on the board it left an imprint! So they started banging and counting. Suddenly Sally exclaimed, 'Stop! We don't need to cover the board. We can simply count how many marks there are in the top row, then how many rows and multiply them together.' After a minute or two Abigail agreed. At this point, however, Jane said 'I don't believe you!' and continued to cover the blackboard with imprints and, on counting them, reached the same result as the others who had used the idea of multiplication. Despite this, Jane still would not accept Sally and Abigail's generalisation. The point of recounting this here was that the teacher had provided the class with many experiences of finding the area of a rectangle by counting using non- standard and standard units. As a result, Sally and Abigail suddenly realised that they had met the idea before when learning about multiplication and thus they appreciated that they could apply the same concept to a similar situation and complete the task more efficiently. Jane, however, obviously needed more experience of counting individual units before she could see the connection.

In illustration 1 the pupils had been given a strategy by the teacher which she thought would be useful but which the pupils were only able to use in a mechanical fashion. In other words, they could use the strategy with little or no understanding of what they were doing. From the children's perspective, they got the right answer, and that is what mattered to them. These 'successful' results, however, gave the teacher the false impression that the pupils understood what they were doing! Our recent work has demonstrated that this 'knowledge' is often enough for children of a particular year group to solve the calculations they are set successfully. However, as soon as they encounter non-standard tasks or more advanced work in another class which requires them to extend their knowledge, they begin to get wrong answers. Such situations pose a real problem for pupils and teachers alike. In the first illustration some of the children developed the misconception that all they had to do was add a zero when they multiplied by 10. For a long time it proved a highly successful strategy for them and, as a result, became well fixed in their minds. As discussed in the previous chapter, such strong associations are hard to shift and, in this case, may require the teacher to go back to first principles and start again. From such strategies pupils get an impression of how mathematics is learnt – 'someone tells me how to do it and then I copy the process'.

In illustration 2 the teacher has a very different approach to learning mathematics. She provided the class with lots of practical examples involving the counting of, for example, matchboxes and books to measure, for example, the area of an A4 sheet of paper and a newspaper. Reflecting on this range of experiments as they were calculating the area of the blackboard, two of the three girls saw the connection between the experiments they had been doing, their current task and multiplication. From this they deduced that there was a general principle they could use in future work on the area of rectangles. What is important is that because they had generated the knowledge for themselves through their own activities, they understood where the formula for the area of a rectangle originated. The reaction of the third girl was interesting since it gave a clear indication that not all pupils learn at the same rate. This third pupil needed more experience of such experiments before she was ready to make the generalisation the other two had made. It would be a great didactical mistake not to provide Jane with further opportunities to gain more experiences.

In our view, therefore, a good curriculum builds up pupils' experiences in a large number of contexts. When appropriate, these experiences can be drawn upon and added to by the teacher to develop a particular concept. Sometimes, as demonstrated below, a single task can be so rich that it can generate sufficient experiences to enable a learner to see the link between the various components and so establish a generalisation for the task.

Figure 6.1 Cubes lying down

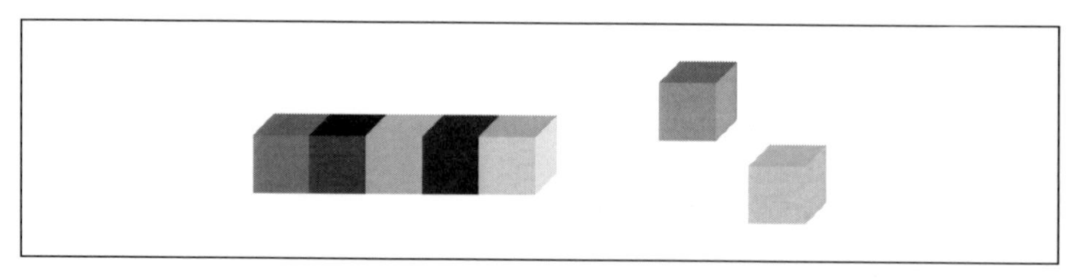

Illustration 3: Patterns. A class of 9–10-year-olds were set the following task:

> Put a cube on your desk. How many cube-faces can you see? Now connect another cube to it so that one of the faces of the new cube is in contact with the desk. How many cube faces can you see now, looking at your shape from all sides? Keep adding cubes, recording your results each time, until you have six cubes linked in a line like a 'train'. How many cube-faces will you be able to see if you have 7, 10, 20 cubes linked together?

Two pupils – Steve and Penny – were of particular interest as they attempted to solve the task. Steve was quite happy to count the cube faces individually and just kept adding cubes one at a time, counting and recording the number of cube faces each time.

Initially Penny counted individual cube faces: for one cube she could see 5 cube faces. She added a second cube, counting the faces in two different ways and then writing '2 cubes joined, 8 faces'. She did this again for three cubes and this time counted 11 cube faces (see Figure 6.1).

Then she stopped and looked at her line of cubes from all angles and suddenly said:

Penny 1 (P1):	I don't think I need to count all the faces each time, I think I can see an easier way to do it.
Teacher 1(T1):	How can you find out if your new way is correct?
P2:	[after quite a pause] I will try my idea and then count them separately to see if I get the same answer.
T2:	Good.
P3:	[Penny then counted the number of cubes she had used, four, and wrote down 14. She then counted all the faces individually as before and her face broke into big smile and she said quietly] It's the same.
T3:	Are you happy your solution works?
P4:	Yes.
T4:	Good. Can you work out then how many cube-faces you would be able to see if you linked 10, 20, 50, or 100 cubes together?
P5:	Yes, ... For 10 it would be 32 ... for 20 cubes 62 faces ... for 50 cubes ... 152 faces and for 100 cubes ... 302.
T5:	It's remarkable that you did it so quickly. How did you do it?
P6:	I saw when I was counting them separately, when I looked from here I could see three equal lines of faces, three on each cube and then there were the two at each end so all I had to do was count the number of cubes linked together and multiply this by 3 and add 2.
T6:	Now, Steve, did you understand what Penny was saying?

Steve 1 (S1): Mmmm I think so!
T7: Go through it again Penny, please.

Penny explained in very simple terms what she had seen and how she got the number of faces.

S2: I get it – you can see three faces on each cube and there are two more each time at the ends.
T8: Good, now I want you to think and also try to see what happens if you build up a 'tower' instead of a 'train'. So start with one on the desk, keep putting cubes on top of it recording the number of faces you can see.

Figure 6.2 Cubes as a tower

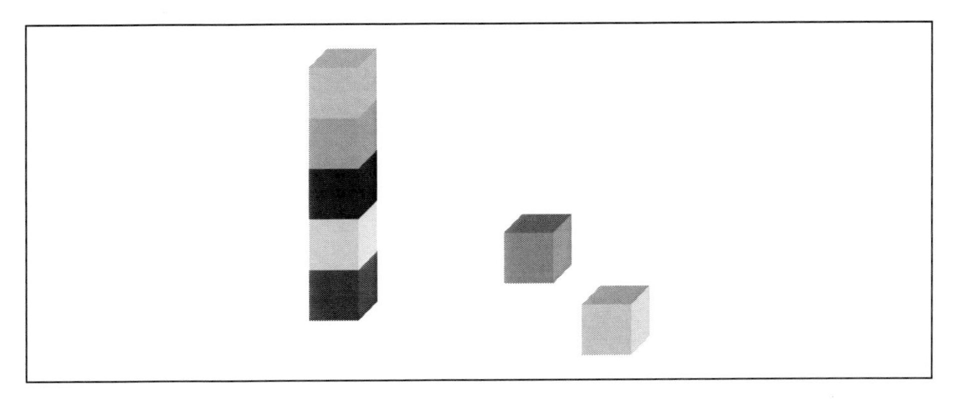

Steve started from one and counted all the faces as he had done in the previous experiment. Penny built a tower of five cubes (see Figure 6.2), carefully studied the result and, adding another cube, announced:

P7: I know how to find the number of faces for any number of cubes and it is different to the last task. I can see four faces on each cube this time and there is one I can see at the end. So for 20 cubes I should be able to see $(20 \times 4) + 1 = 81$. I checked with five and six cubes and I was right!
T9: What about you Steve?
S3: I was counting them all but when I heard Penny talking to you I had another look at the 'tower' and saw what she meant. Is it right if I say that I could see 401 faces if my tower was 100 cubes high?
T10: Well done! You have followed Penny's reasoning very well. Now I have one final task for you. I have put out four piles of matchsticks. Again I want you to find out how many matchsticks I would have to put out to make the next two piles. Can you make patterns so that you could easily find out how many matchsticks you would need for the 10th, and 20th piles? Work together if you wish.
S4: We shall have to count how many there are in the piles Mr T has given us.
P8: Yes. There are ... four in the first pile ... 7 in the next ... 10 in the 3rd and are there 13 in the last Steve? (See Figure 6.3.)
S5: [Steve counts the sticks] Yes, how did you know?
P9: Well, look! The numbers are 4, 7, 10 there are three more in each group so add 3 to 10 and there should be 13 in the fourth group.

Figure 6.3 Three piles of matchsticks

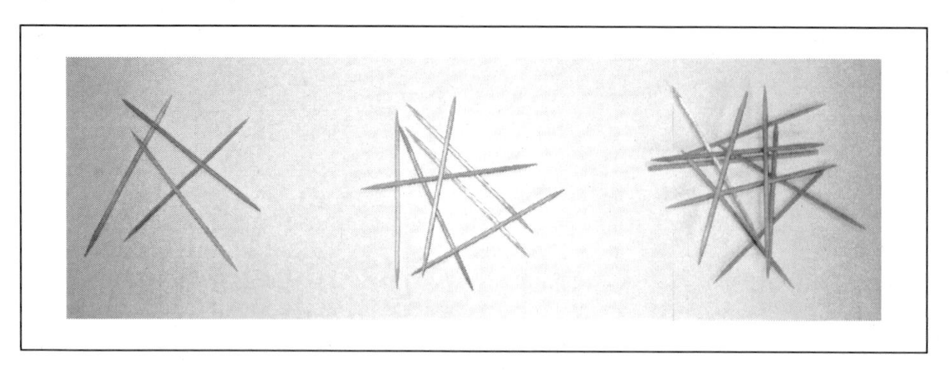

S6: Aren't you clever! So the next will be 16 and the one after that 19! Let's see if we can make a pattern with the sticks. As you said we keep adding three more so let's make a triangle or several triangles. See with the 13 I can make four triangles with one left over.

P10: We should start at the beginning and not half way along. We'll try your idea with all the piles: four sticks makes 1 triangle and one left over.

S7: 7 sticks makes 2 triangles and one left over and see 10 makes 3 triangles and 1 left over. (See Figure 6.4.)

P11: Let's draw what we have found and write it down.

Figure 6.4 Matchsticks in pattern

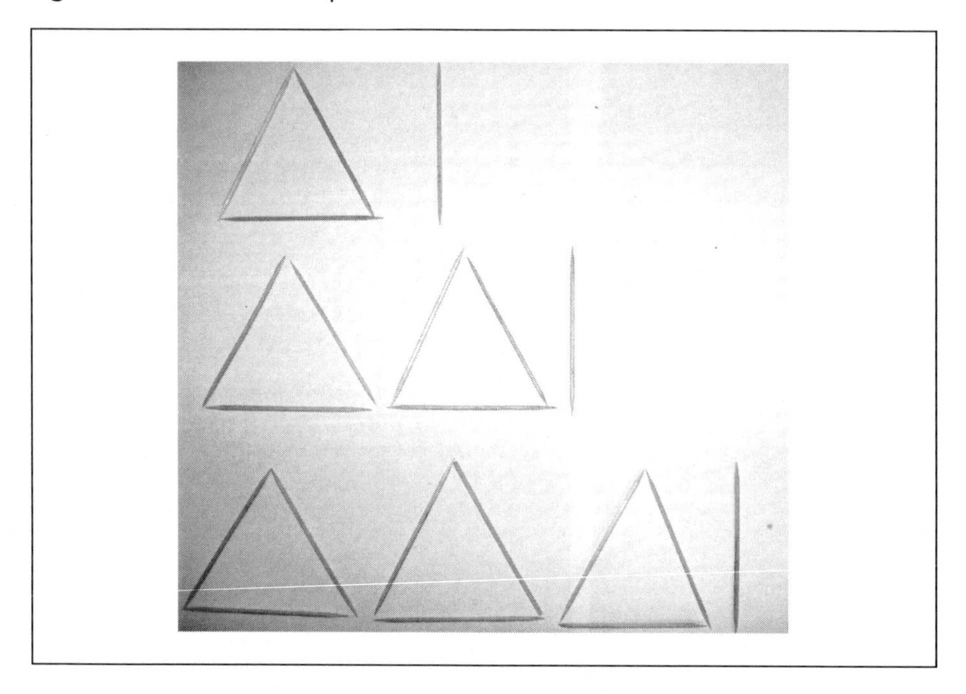

So Steve drew pictures of the shapes they had made and Penny made a chart, as shown in Table 6.1.

Table 6.1 Penny's chart of the patterns made from the matchsticks

Number of piles	Number of triangles	Number left
1	1	1
2	2	1
3	3	1
4	4	1
5	5	1

S8 and P12: I can see a pattern!

P13: What pattern do you see, Steve?

S9: The number of triangles gets bigger by one in each group and there is always one left over!

P14: I agree but my table shows that there is the same number of triangles as the number of the pile. The teacher asked us to find how many in the 10th and 20th piles and now I can tell him.

S10: So could I. I would have 10 triangles and one more in the 10th pile – mmm ... 31 sticks.

P15: I agree and there would be twenty lots of three plus one, sixty-one, in the 20th pile. So we could find out how many sticks are needed for any pile. All we need to know is which pile, multiply this by 3 and add 1 and that's how many sticks we need. Do you agree, Steve?

S11: I think so!

Comments

The two pupils were clearly motivated by the tasks the teacher had set them. Asking them to work together helped Steve. By working with Penny, he was able to gain insight into how she solved the tasks.

Note the way in which the first task encouraged the children to experiment by getting them to record the number of cube faces they could see each time they connected a new cube. This provided several opportunities designed to encourage them to see a pattern in their results. Steve needed a considerable number of these experiences before any links were established in his mind and, in fact, in the first example he continued experimenting, adding a cube each time until the teacher talked to Penny.

Penny, on the other hand, after recording only three results in the first task saw a link between them – a pattern – which was reinforced by her visual perception of the task (P1). She was able to tell the teacher therefore how many cube faces could be seen when four cubes were in a line. To check her new mathematical knowledge she counted the faces she could see. Getting the same answer from both methods gave her the confidence to say how many cube faces would be on a 'train' of any number of cubes (P5). That is, she was able to generalise her thoughts to describe her solution of the task (P6). We would not expect pupils of this age to take this any further, but by the time some children are 10+ years they would be able to express Penny's conclusion in a symbolic form, namely $3n + 2$, where n is the number of cubes joined together.

On realising that Steve had not followed Penny's reasoning fully, the teacher then used a clever ploy, asking Penny to go over her conclusions again. As a result, Steve was able to make his own abstract statement (S2).

The teacher then gave the children a very similar task to see whether they would use the ideas they had just developed and thus consolidate the way of making links and generalisations through working with different experiences. Penny did not need many examples before she could establish the link between them. She used the model she had formed from the previous task by immediately building a tower of five cubes, studying it from all angles and counting the number of cube faces she could see. She made a deduction, but added a sixth cube to check whether she was correct. She then expressed her result as *'I can see four faces on each cube this time and there is one I can see at the end'* after which she used this knowledge to calculate the number of faces which could be seen on a tower of cubes 20 high (P6).

Again Penny used a small number of experiences to find a general expression by which she could calculate the number of cube faces visible for any tower of cubes. She was working with the five-cube tower as a general case for all other towers and verified this with the concrete example of six cubes. Steve obviously listened carefully to what was going on. He did several experiments, recording his findings each time but, by the time Penny was talking to the teacher, still had not made any connections or seen a pattern in his results. Penny's explanation of her work obviously resonated with what Steve had done previously and he was then able to see a connection between his results and Penny's generalisation. This gave him the confidence to state how many faces he would be able to see if he had a tower 100 cubes high (S3). Thus, through his experiences and working with and listening to Penny, Steve had assimilated new knowledge with understanding.

The teacher then set them a final task using different material to consolidate the children's knowledge and to see if they could apply it in a different situation.

Steve quickly realised that he had to count the matchsticks to get an idea of the sequence of numbers (S4). He counted all four piles carefully but Penny, after counting the first three piles, saw the pattern in the sequence and told Steve there would be 13 in the fourth group (P7). He was surprised at this and Penny had to show him that the difference between each of the first pairs in the sequence was 3, hence the fourth group should contain $10 + 3 = 13$ matchsticks (P8). Steve made a rather scornful remark but then used Penny's idea to give the correct solutions for the fith and sixth terms of the sequence. Again he was learning with Penny's help (S6).

Steve rapidly wanted to move on from Penny's insightful contribution and suggested they should arrange their matchsticks in patterns as – apparently almost without thinking – he had arranged the fourth group of matchsticks into four triangles and there was a match left (S6).

Penny wanted to be more methodical, starting with the first term and recording what she was doing. Steve drew his patterns and Penny recorded them in her chart after making them. Simultaneously, they both announced they could see the pattern (S8 and P11). This was a big step for Steve: previously he needed to hear Penny's explanation before committing himself, but now he had the confidence to make his own statement.

Steve was able to generalise by expressing his 'geometric model' in terms of the number of triangles he could make and always having one left over (S9). Penny added that her chart showed that the number of triangles was the same as the number of the term (P13). So they both agreed they could find the 10th and 20th number of matchsticks.

Finally, Penny gave a definition of the generalised idea arising from their experiment: *'All we need to know is the number of the pile, multiply this by 3 and add 1 and that's how many sticks we need'* (P15).

The teacher in our illustration realised that Steve needed more examples to develop his understanding and so gave a sequence of tasks: the first two were similar, so Steve could apply what he had learned from the first task to the second, but the third task was slightly different, demanding the recognition of a pattern. Working with Penny also helped Steve, illustrating that sometimes it can be highly effective to acquire knowledge from another pupil, thus gaining second-hand knowledge from a 'non-authoritarian' person rather than a person of 'authority' (Hejný and Littler, 2007). It also has the advantage of freeing up the teacher to work with others who might not readily benefit from working with a peer.

We now briefly explain Hejný's theory, referring to our illustrations which demonstrate three of the five stages of the theory. Like all such discussions, we get a bit technical from time to time, but don't be put off as it will be worth the effort!

Hejný's theory of generic models

Using Hejný's theory of generic models (TGM) (Hejný and Kratochvílová, 2005), the learning process can be broken down into the following stages:

> **individual experiences** → generalisation → **generic model** → abstraction → **abstract knowledge**

TGM comprises two distinct levels. At the first level the pupil gains individual experiences through activities, which initially are not linked but stored in their memory as isolated models for future knowledge. Then these individual experiences begin to be linked together in the child's mind and he/she starts to use one particular case to represent more individual experiences. We say that the pupil created a *generic model* by the *mental process of generalisation.*

Pause for thought

Thinking back to Penny and Steve, can you apply Hejný's theory to how their mathematical understanding developed?

At the second level the generic model loses its basis in practical mathematics and, through the mental activity of abstraction and by changing the language used, it changes into abstract knowledge which can then be applied to all such situations.

You can see that our two pupils in illustration 3 have developed the generic model in all three tasks. In other words, having undertaken some well-planned, sequential tasks, Penny and Steve were able to generalise their learning. Later they will be able to progress to the next stage of the learning process and develop the abstract knowledge. At the moment they can only express the relationship in words.

We hope our illustrations above emphasise the theoretical model and show the pupils' learning process. The number of experiences needed to create a generic model differs from pupil to pupil and from task to task. In cases where part of the stage is shortened, rushed, curtailed or skipped, the pupils may create mechanical knowledge which could be a source of misconception in their minds. This is what happened in illustration 1 above where some children began to add zeros indiscriminately when multiplying by 10.

An application of Hejný's theory in adult learning

Hejný's learning theory applies not only to pupils in school but also to all learning at all ages. Using a task similar to the one found in Chapter 2 (Figure 2.8), we will now illustrate each stage of the theory to help you develop your teaching strategies to enable children gain a deeper mathematical understanding and, in so doing, lessen the chance of them developing misconceptions.

Consider the following task:

> *Task A*
>
> Given four different digits A, B, C, D such that A is less than B and B is less than C and C is less than D (i.e. $A < B < C < D$), develop a strategy to create two 2-digit numbers such that when they are subtracted the result is the least possible positive number. In other words, develop a strategy for distributing the four digits to gain the smallest possible positive result in the format:
>
> [__ __ – __ __ = ?]

Unlike the exercise presented in Chapter 2, the task above requires you to develop a general – as opposed to a particular – solution for any four different digits (0 to 9) you choose. This means you should endeavour to express the solution in abstract form. For instance, your solution might be:

> Put the two smallest numbers in the tens places and the two largest numbers in the units places, both pairs in reverse order of value. When you subtract these two 2-digit numbers you get the smallest positive result.

Challenge

Try task A for yourself before reading on.

Having tried the task, you will now have more insight into the processes we are advocating for pupils. Now compare your solution with the one we present in four stages below, so that you can see the development following Hejný's theory.

First stage

1.1 *We chose four digits at random, for example 1, 2, 5, 7. We then randomly constructed two 2-digit numbers from the four and found their difference, just to get a feeling of what the task involved, for instance 57 – 21 = 36.*

By doing this we gained insight into the situation and, as a result, produced our first isolated model.

1.2 *We could see that this difference was quite big. We recalled the strategy we used for determining the greatest sum of two 2-digit numbers and used this strategy to create two new numbers 71 – 52 = 19, hence getting a much smaller difference.*

In other words, we came up with another isolated model. Our next step was to look for some connection between these two isolated models.

1.3 *We felt the difference was getting reasonably small, but we tried another pair with the aim of making the 2-digit numbers smaller: 51 – 27 = 24. We then attempted to make the numbers even smaller and created 27 – 15 = 12, the smallest result so far.*

By enriching our collection of isolated models it became easier to concentrate on the links between them.

1.4 *We continued with this idea and created the numbers 25 – 17 = 8 which could be the smallest possible result. This led us to articulate hypothesis 1:*

> *Given four digits, order the numbers from smallest to largest, take the first two digits and put them in the tens places in reverse order, then take the second two digits and put them in the units places in that order.*

So, by working through several examples we were able to produce a general statement which, if it is correct, could be applied to any four digits in the form of an instruction for creating the desired two numbers. Note that the wording does not define specific digits and therefore any digits could replace those we used. In other words, our hypothesis is a generic model in terms of Hejný's theory. Should we wish to, we can also present the generic model in a more abstract form expressing the strategy so that it represented all possibilities: $(B \times 10 + C) - (A \times 10 + D)$.

Although we were pleased with hypothesis 1 we were aware that it might not be correct and therefore we moved onto the next stage to check it.

Second stage

2.1 *To verify hypothesis 1 we decided to use different digits: 1, 2, 3, 4. If we use hypothesis 1 on these digits we get 23 – 14 = 9 which suggests that our hypothesis might be sound. Having said that, we thought it important to check that 9 is the smallest possible result.*

2.2 *We used a table to list all the permutations of the four digits which gave a positive result, to make the values easier to see.*

Table 6.2 provides an overview of all the possibilities. Looking at the isolated models you can see the links between them, and it was at this stage we realised that our model required further work before it could be generalised.

Table 6.2 The results of subtracting a range of two-digit numbers

				Diff.
23	–	14	=	9
24	–	13	=	11
31	–	24	=	7
32	–	14	=	18
32	–	12	=	20
34	–	21	=	13
41	–	23	=	18
41	–	32	=	9
42	–	13	=	29
42	–	31	=	11
43	–	12	=	31
43	–	21	=	22

2.3 *More specifically, two of the answers surprised us. We found another way to get 9 but, more pertinently, we noted an even smaller result, 7. In other words, hypothesis 1 broke down when we changed the digits. This led to our second hypothesis:*

> *The smallest difference will be given by taking the two middle digits and putting them in the tens places in reverse order and then the outer two digits in the units places in their order.*

So, summarising what we know: hypothesis 1 related only to a limited set of quadruples of digits, and it would appear that hypothesis 2 is another generic model which, as a minimum, applies for the special cases of consecutive digits, in this case 1, 2, 3 and 4.

2.4 *We then checked whether hypothesis 2 held for a different set of four consecutive digits (e.g. 6, 7, 8, 9). It did and, again, we were able to produce the hypothesis in an abstract form: $(C \times 10 + A) - (B \times 10 + D)$ will produce the smallest positive result in the case of four consecutive digits.*

So now we have two formulated hypotheses, but both are only partial as they each relate to limited sets of digits. The next step therefore is to find a final general strategy (the required abstract knowledge) which would be the generalisation of several partial hypotheses.

2.5 *Looking back at Table 6.2, what do the two cases giving the same result in the table have in common and how do the two cases giving the result 9 differ from the one giving the result 7? Close examination reveals that the difference between the numbers played a role. In the cases where the difference is 9, the digits in the tens and units places are consecutive. This was not the case when the result was 7: here the digits in the tens places were consecutive and those in the units places are not. The difference between these latter two digits is the biggest possible.*

Recognising the importance of the distances[1] between numbers in a quadruple further enriched our understanding of the situation and enabled us to progress to a third stage.

Third stage

3.1 *Our next step was to test another set of four digits where the distance between each two neighbouring numbers was 2, e.g. 1, 3, 5, 7. We found that hypothesis 2 worked for all these cases. We also made another discovery – the difference of the two 2-digit numbers related to the distance between the neighbouring digits. (The study of this relationship is another task which, if you are an enthusiast, you might like to undertake.)*

3.2 *Next we tested a range of quadruples – 1,2,3,7 and 1,4,5,6 or 1,3,5,9 and 1,5,7,9 – and concluded that these cases extended the range for which hypothesis 2 was true.*

Fourth stage

4.1 *We tested quadruples of the following types: 1, 2, 4, 7; 1, 3, 4, 7; 1, 4, 8, 9 and 1, 2, 8, 9. Four more sets of digits where the distances between the neighbouring digits are different from those in 3.2. Our conclusion was hypothesis 3:*

> *Given four digits, order them from smallest to largest and create two 2-digit numbers as follows: take the two numbers with distance 1. Put them in the tens places in reverse order. Take the remaining two numbers and put them in the units places in that order. Subtract these two 2-digit numbers and you get the smallest possible result. If there are two possible pairs of numbers with distance 1, there are two solutions to the task.*

4.2 *So now we can generalise this hypothesis into general strategy:*

> *If four digits A, B, C, D are given and their numbers ordered from smallest to largest, create two 2-digit numbers as follows: take the two numbers with the smallest distance and put them in the tens places in reverse order. If there are two possibilities,*

1 By distance between two natural numbers *a* and *b* we mean $|a - b|$.

> *choose the one for which the remaining two numbers have bigger distance. Take the remaining two digits and put them in the units places in that order. On subtracting the two 2-digit numbers, the result will be the smallest possible. If there are two possibilities for the above choice, then there will be two solutions.*[2]

You may have noticed that stages 3 and 4 were briefer that 1 and 2. This is because we were able to build on our findings from the earlier stages. As many primary pupils are encouraged to do, we expressed the abstract hypotheses (abstract knowledge) in words first. Later we presented it in abstract symbolic form (1.4 and 2.4). In essence we worked with generic models and developed them through abstraction to a final strategy to present abstract knowledge. Thus we demonstrated Hejný's TGM.

So let us return to the challenge we set you earlier in the chapter: how did your strategies compare to ours? Interestingly, when we gave the task to an experienced mathematician he, like us, did not give us an immediate solution but went through the process of using – albeit very few – generic models. As someone new to the task you may have used rather more. That is entirely understandable. Each step you took would have revealed something of your thinking and, we hope, your developing understanding. More generally, we would conclude that if you as a teacher are aware of this process of gaining knowledge you are more able both to understand pupils' mistakes and to design their learning process. We will come back to this idea later on when we discuss pupils' mistakes in the solution to a particular task.

Tasks for pupils

It is important to motivate children. We have found that the most successful tasks are often non-standard (i.e. not typically found in a textbook) but set in a context the children recognise. They are also challenging but extend children's knowledge in an interesting and non-threatening way.

Challenge

(a) Given the six-digit number 273194, cross out one digit so that the remaining five-digit number is the *largest* number without reordering any of the remaining five digits.

(b) Given the same six-digit number 273194, cross out one digit so that the remaining five digit number is the *smallest* number without reordering any of the remaining five digits.

Hopefully you found the above task stimulating and perhaps a bit more challenging that you had anticipated. It can be altered to suit people of different

2 For example, for 2, 4, 7, 9 the two solutions are 47 – 29 = 18 and 92 – 74 = 18. In the case of 2, 4, 6, 9 from the two possible pairs, 2, 4, and 4, 6 which have the same distance we choose 4, 6 since the two remaining digits, 2, 9 have a bigger distance than the other remaining pair 6, 9.

age groups. So, for a year 3 class you might use a three-digit number such as 594. If you cross out the 5, you will get the largest two-digit number, 94, and if you cross out the 9 you get the smallest 2-digit number, 54.

The task can be described more generally as:

> *Task B*
>
> *Given an n-digit number (n being dependent on the age/ability of the pupil), strike out a digit so that what is left is the largest possible (n – 1)-digit number. The digits in the original number must not be reordered. Starting again with the original number, strike out a digit to make the smallest possible (n – 1)-digit number. Thus, n in the case of 594 above would be 3 with the largest 2-digit number resulting being 94 and the smallest 54.*

When designing such tasks our objectives were to:

(i) give us insight into the pupils' knowledge and understanding of place value. So, for example, in the case of 594 do children consider *place value* (as we would hope!) or opt for crossing out the 'largest' digit (i.e. 9) when endeavouring to create the smallest two-digit number?

(ii) see if there were common misconceptions across the four countries of the project. If not, what could we learn, for example, about the different teaching methods used?

(iii) analyse the misconceptions to determine their origin. This included looking for patterns in the children's responses to see if, for instance, they responded correctly to all the tasks they were given excepting those involving zero(s).

(iv) provide ideas for re-education/education to eradicate the misconception(s).

Figure 6.5 Predicted strategies for completing task B

- Strategy 1: Cross out the 'smallest' digit to get the largest number.
- Strategy 2: Cross out the 'largest' digit to get the smallest number.
- Strategy 3: Cross out the right-hand digit to get the largest number.
- Strategy 4: Cross out the left-hand digit to get the smallest number.
- Strategy 5: Cross out the zero to get the largest number.

Our experience suggested that the strategies presented in Figure 6.5 were the most likely, so we devised 3-, 4-, 5- and 6- digit numbers such that, if the pupils used one of these strategies, they would not get the correct answer. Table 6.3

Table 6.3 A range of solutions when finding the smallest and largest numbers

	Correct solution	Smallest/largest digit – strategies 1 and 2	Right/left digit – strategies 3 and 4	Zero, largest smallest – strategy 5
213				
Smallest	13	21	13	
Largest	23	23	21	
120				
Smallest	10	10	20	12
Largest	20	12	12	12

details the range of possible results if pupils started with the numbers 213 and 120 and used the strategies outlined in Figure 6.5. Table 6.4 provides the actual results for three groups of 6–7-year-olds (children).

Interestingly, our analysis of the pupils' solutions showed that all the strategies previously discussed were used by this age range of children, so we did not have to go further than year 1 to answer our question 'How early in the pupils' schooling were the strategies described above established?'.

We also noted that fewer than 50% of the 58 pupils got the two parts of the task correct. There were 9 pupils who used the twin strategies 'cross out the left-hand/right-hand digits for the smallest/largest numbers, respectively'. No pupil who used 'cross out the smallest digit to get the biggest number' used 'cross out the biggest digit to get the smallest number'. What do you deduce from this?

Eleven of the pupils used the twin strategies 'cross out the left- and right-hand digit to get the smallest and largest numbers, respectively'. Zero caused many pupils difficulties since many children appeared unsure of its function. They probably have been told that 'zero is a place holder' or that 3 – 3 'is nothing' which is then written as '0'. Other phenomena connected with zero arise with older pupils (see Chapter 0).

Closer examination of the results also revealed illuminating material about the three different classes. For example, in one of the three groups, we suspect that the children were given some guidance on how to find the smallest number for 213 since the whole group gave the same incorrect solution: cross out the '3' and then reverse the remaining digits '21' to get the number 12. Interestingly their solutions for the other tasks were not significantly different from the other two groups.

Also we observed that in one class the most common misconception for finding smallest number for 213 was crossing out the '3'– the largest digit. Another group had different – contradictory – misconceptions for finding the largest number in 213 as the same number of pupils crossed out the right digit, as crossed out the left one. In this class more pupils crossed out the right digit, '0', in 120 to get the largest number than got the correct answer. A number of pupils rearranged the digits if their strategy did not seem to give the expected result, even though we know that they were told at the start of the tasks not to alter the order of the digits.

Pause for thought

When you tried the task with your class, did you find a significant number of pupils producing the same incorrect solution and yet otherwise generally performing well? How might you account for it?

We are sure some of these strategies are developed in the pupils' minds when they misinterpret something they have heard and/or they do not fully understand place value. Many children have the concept of the value of a number and understand the value of the places in the decimal system for natural numbers,

Table 6.4 The results of children's work on task B expressed as percentages

Numbers → Strategies ↓	**Making largest number from 213**	**Making smallest number from 213**	**Making largest number from 120**	**Making smallest number from 120**
1: Cross out the 'smallest' digit to get the largest number	**47**			
2: Cross out the 'largest' digit to get the smallest number		12		**57**
3a: Cross out the right-hand digit to get the largest number	17		21*	
3b: Cross out the left-hand digit to get the largest number	21		**72**	
4a: Cross out the left-hand digit to get the smallest number		**41**		
4b: Cross out the right-hand digit to get the smallest number				33*
5: Cross out the zero to get the largest number			21*	33*

*Signifies that, in the case of 120, strategy 5 produced the same result as another strategy and hence the figures have been duplicated as it is unknown which of the two strategies the children used.

but they do not connect the two concepts (Ashlock, 2002). This is when the importance of structural apparatus is paramount. Material such as Dienes multi-base arithmetic blocks is invaluable for linking these two concepts together.

We know from talking to pupils that some have the misconception that if they take out the largest digit then what will be left is the smallest number, and vice versa. The idea behind this being that if you take out the largest digit then only smaller digits will be left and hence the smallest number. This does not take into account the positional value of the digit. We only need to look at Table 6.3 to see that taking out the largest digit in 213 does not give the smallest number and taking out the smallest digit in 120 does not give the largest number.

Pause for thought

Do you think that the difficulties children experience are partly due the modern tendency to digitise numbers? The 125 trains in the UK, for example, are always spoken of as the 'one, two, five trains' and never the 'one hundred and twenty five trains'.

When the project teachers tried the task with 8–10-year-olds using four-digit numbers, we found that all the strategies used by the younger children reappeared, with crossing out of the largest and smallest digits being the dominant misconception of the pupils. We were also able to develop new insights into the children's thinking, for some of those who said they were crossing out the smallest digit to get the largest number in 2109 and 9120, crossed out the '1' and not the '0', getting 209 and 920 respectively. This gives the incorrect answer for the first number and a correct answer for the wrong reason in the second. It seems likely, in our view, that they did not see zero as a digit having value but purely as a place holder in the denary place value structure.

This brings us to another important point: whenever possible, try to look at a pupil's working to identify how he/she has got to the answer. It is all too easy to whiz through a page of marking only checking the answers as there are so many other demands on your time but, if you can take a little longer, we are confident you will be able to identify misconceptions – such as the one underlying 920 above – at an early stage and take appropriate action (see below). Such detailed examination of children's work is especially important where pupils were not consistent in their strategies for solving the tasks. Some pupils started with what appeared to be a discernible strategy and something occurred during their work which cast doubt on the method they were using and they changed their strategies, not necessarily to the correct ones.

Challenge

We found that when some 12-year-olds applied the strategy 'cross out the smallest digit to get the biggest number and cross out the biggest digit to get the smallest number' with the number 352091, they produced answers of 35291 and 35201. Why do you think this might have come about? How might you help them overcome their misconceptions?

Several 12-year-olds crossed out the '1' rather than the '0' when working with numbers such as 352091, making us question whether they really understood the concept of zero despite the fact that they were generally able to use it in calculations. Another interpretation might be, however, that they do not realise that 0 – along with the numbers 1–9 – is a digit.

Another strategy used was 'cross out the right-hand digit to get the biggest number and the left-hand digit to get the smallest digit', giving the answers 35209 and 52091, respectively. This showed, as before, that the pupils were doing what they thought was 'mathematically' correct and not looking at their answers to see whether or not they were sensible, since the smallest number was considerably larger than the largest number! Although we encourage children from a very early age to check their answers, such responses remind us that, for some pupils, it is a never-ending task.

In brief, therefore, we found that exactly the same misconceptions occurred with those classes working with five-digit numbers as occurred with younger children. What is even more astonishing is that some year 10 pupils working with six-digit numbers appeared to have the same misconceptions as 6-year-olds, suggesting that some mathematical misunderstandings are very deep-seated and probably start to germinate in the earliest years of schooling. This is not, in any way, to cast blame on those of you who teach younger children, but rather it broadens our insight into some of the complexities of teaching and learning an abstract and complicated subject such as mathematics.

Finally, in one country a worksheet was devised which gave examples of 3-, 4-, 5- and 6-digit numbers to year 4 pupils. This was particularly interesting since we were able to follow the pupils' thinking to see if they applied consistent methods throughout, regardless of the size of the numbers presented. When they were inconsistent we suspect that some found that a strategy they had used with the smaller numbers was no longer adequate and so they looked more closely at what was required of them. Moreover, several pupils showed clearly a phenomenon identified by Mason and Burton (1982) – that they were thinking mechanically until a problem was encountered, then they really started to think for themselves.

Misconceptions – what were they and how did they happen?

The important results of our analysis were as follows:

- The strategies listed earlier in this chapter were prevalent in year 1 and were found in every grade up to year 10.
- Most pupils were inconsistent in the strategies they used to solve the problems both across tasks and within tasks. This would suggest that these pupils considered each task individually. Not many pupils used both of the twin strategies – smallest/largest digit or right/left-hand digit to solve one task.
- Pupils did not check to see if their answers were sensible.

- Many pupils did not connect the cardinal value of the digit with the place value where it was situated.
- Very few pupils used the strategy of 'list all' and cross out the digits in turn to determine smallest and largest number.

The pupils really thought that their strategies would give them the correct answers and so did not attempt to consider the solution of the tasks from first principles. From the point of view of the learning processes described above, the pupils were at the level of getting isolated models and were not focusing at a generic level.

As discussed before, therefore, it is likely that mathematical misconceptions occur because pupils are not given sufficient experiences of a concept before the class moves on to solving tasks involving the concept. Thus the child needing more familiarisation has to be satisfied with second-hand knowledge usually supplied by the teacher or a fellow pupil, leading to mechanical knowledge or misinterpretation of information without understanding. Hence, statements – such as 'you always subtract the smaller number from the larger' – which have been made for given circumstances are interpreted as being relevant for all situations and a misconception is born.

Much of this relates to Fischbein's (1987) ideas on intuition (see Chapter 3). For instance, when pupils are learning the beginnings of number, 1 is seen as being the smallest quantity and 9 as the largest number, therefore if I have a multi-digit number and strike out the '1' intuitively I will be left with the largest number (without considering place value), and similarly if I cross out a '9' then what remains must be the smallest number. A similar argument could be made for crossing out the right- and left-hand digits: the units column has the smallest value in place value, so if I cross that number out I will be left with the largest number. In a six-digit number the left-hand digit represents hundreds of thousands, so if I cross out that digit the number cannot be bigger than tens of thousands. These intuitive extrapolations from not fully understood basic concepts lead to the misconceptions we have discussed above. Once firmly fixed in the pupils' mind they are difficult to eradicate without going back to first principles.

Re-education

Incorrect strategies which yield the correct answers in certain cases are some of the most difficult misconceptions to eradicate, simply because they are usually deeply ingrained in the pupils' memories.

The basic problems are: how do we show pupils that their intuitive extrapolations of basic concepts do not work? How do we make our pupils more aware of the value of digits when they are in a multi-digit number? How do we get our pupils to check their work to see whether or not their answers are sensible? (With regard to the last of these: some of our teachers were particularly frustrated when many pupils gave answers where the smallest number in the tasks above was larger than the so-called 'largest number'!)

To start the remediation for these misconceptions, we believe it works best if children are confronted with conflicting situations which clearly show them that their strategies do not work in all cases. Such tasks can be tricky to devise, so it is important to try solving them beforehand and choose tasks which in some circumstances work for one strategy and not for another. On setting the tasks it is crucial that you do not tell the child that his/her strategies do not work. In part this is because their preferred strategy is likely to be too well established in the child's memory, and in part because – in the light of the strategy yielding some correct answers in the past – he/she is unlikely to believe you. In other words, it is important for children to discover for themselves that a particular method may not work in all situations. Giving plenty of examples is generally the most effective way for pupils to come to this conclusion themselves (Hejný's TGM).

By way of an example, let us go back to first principles and, using the tasks discussed above, start with two-or three-digit numbers and give a series of simple numbers which show that the children's intuitive ideas are false. For instance if we cross out the smallest digit in the number 231 we do not get the largest number. Rather we need to cross out the 2 to get the largest, 31. Here we also see that by crossing out the left-hand digit we do not get the smallest number, just the reverse in fact. Again, with the three digits 427, crossing out the largest digit does not give the smallest number, nor does the strategy of crossing out the right-hand digit. Moreover, to get the smallest number you need to cross out the 4 to get 27 and to get the largest number cross out the 2.

When working in this way it is useful to get the pupil to write down their first response. So if given 427, and they suggest crossing out the '7', ask them to write the resulting 42. Then invite them to express this in terms of tens and units, $4 \times 10 + 2$, so that they link a digit with a place value. If, on searching for the smallest number, they arrive at the correct answer of 27 again, write it in the form $2 \times 10 + 7$, so that they can compare the two values. We also advise that the pupils use the 'list all' strategy (as we do in Table 6.5) so that they are able to see the smallest and largest numbers clearly and note which digits have been removed to get the correct solution. Once children appear confident with 3-digit numbers we suggest you present them with similar tasks involving 4-, 5- and even 6-digit numbers if appropriate, to ensure that their misconceptions have been eradicated. With older children, apart from the linking of digit values with place values, it is good to get them to cross out the digits in order so that they can see more clearly what are the values of all the five-digit numbers which are the result of crossing out each digit in turn from the original six-digit number. For example, with the number 352091 they could create Table 6.5. They will see that only one of the 'solutions' starts with fifty thousand.

Table 6.5 Comparison of different solutions using the systematic crossing-out technique

352091 cross out 1st number gives	52091
352091 cross out 2nd number gives	32091
352091 cross out 3rd number gives	35091
352091 cross out 4th number gives	35291
352091 cross out 5th number gives	35201
352091 cross out 6th number gives	35209

So, to summarise, we have observed the following:

- Going back to first principles tackles the difficult task of removing these false strategies from the pupils' minds.
- Getting the children to write down the value of the resulting numbers, both wrong and right, helps to link the digit value with its place value.
- Getting the pupils to cross out all the digits in turn will make them look at the 'solutions' more closely and so relate their final answers to the actual question.

Further reading

Hejný, M. and Littler G.H. (eds) (2007) Transmissive and constructivist approaches to teaching. In *Creative Teaching in Mathematics.* Prague: Charles University.

This is the introductory chapter in a book specially prepared for European in-service courses. Teachers from many countries have found its examples and exposition very useful in helping them to understand the difference between the types of teaching strategies and how they influence children's learning.

References

Ashlock, R.B. (2002) *Error Patterns in Computation: Using Error Patterns to Improve Instruction.* Upper Saddle River, NJ: Merrill Prentice Hall.

Fischbein, E. (1987) Intuitions and schemata in mathematical reasoning. *Educational Studies in Mathematics,* 38, 11–50.

Hejný, M. and Kratochvílová, J. (2005) From experience, through generic models to abstract knowledge. Paper presented to CERME 4, Working Group 3, Spain. http://cerme4.crm.es/.

Hejný, M. and Littler G.H. (eds) (2007) Transmissive and constructivist approaches to teaching. In *Creative Teaching in Mathematics.* Prague: Charles University.

Mason, J. and Burton, L. (1982) *Thinking Mathematically.* London: Addison-Wesley.

Everyday numbers under a mathematical magnifying glass

Carlo Marchini and Paola Vighi

Introduction

In this chapter we explain the basic elements of number from a more mathematical perspective than previously discussed, picking up threads from previous chapters. Do not let this put you off! The topics we consider are not intended for everyday use in school, but rather to provide you with a deeper understanding of number so that you can create new connections between the key mathematical concepts and use them for teaching your pupils more effectively. For example, we know only too well the awkward feeling of having to teach something that it is not completely clear to us. It can be particularly challenging when we reach this conclusion half way through a lesson as children are often quick to spot when we are in difficulty. Here we will simply dip our toes in the water of the great ocean of mathematics, but we hope you will find it helpful rather than be put off the subject for life!

We are aware that for non-specialists, one of the most challenging aspects of mathematics is all the jargon in which formalism lords it, but sometimes it is salutary to remember that such formalism is the result of thousands of years of searching for clarity and conciseness. Therefore, acquiring the jargon can be seen as another mountain to climb on to the shoulders of the giants preceding us!

Before we proceed, there is another important issue to consider. Research shows that a person's cultural and social background can significantly influence their mathematical performance.

The most important mathematical topics are presented here on a light grey background so that readers who are better informed – or pressed for time – can skip sections and concentrate on the key concepts.

Number systems

Pause for thought

Imagine that our culture had evolved without digits, but only with letters. How do you think it might operate?

Numbers are useful for labelling houses, buses, telephones and so on, but they are not essential and we could use other signs – such as letters – instead. Numbers are also used as labels, and these can convey information about order: for example, the house numbered 33 is generally next to number 31, although the same cannot be said for buses. Harry Potter's platform 9¾ is between lines 9 and 10!

A strength of number is the fact that with a small alphabet (ten signs, the *digits)* we can write an infinity of different 'words' or, more correctly, *numerals*. Letters share with numbers the fact that order can play an important role. Thus, the convention in English is that *a* comes before *d* in the alphabet. For the digits we have a universally accepted (natural) order which reflects the cardinality. Therefore, we can state that 2 comes before 4 when the numbers are ordered smallest to largest, and we can also express this fact by saying that 2 is smaller than 4, 2 is less than 4, 2 is fewer than 4 and so on. Numbers hold another trump card. We can manipulate them by the means of suitable operations – such as addition and multiplication – and algorithms. In other words, they can be used in many other ways than simply as 'labels'.

In this chapter we will illustrate how we can pass from the number system you are most familiar with to another richer one. The latter is the result of considerable mathematical research which originated in Cambridge at the beginning of the nineteenth century, by the so-called Analytical Society, the outstanding members of which were Woodhouse, Babbage, Peacock and Herschel (Mangione and Bozzi, 1993).

Challenge

Stop reading for a while, and write down which kinds of numbers you know and how you might use them.

If you are reading this with someone else, do the task singly and then discuss what you have done.

Then, resume reading and check your statements, comparing them with ours.

Properties of operations in number systems

Let us begin by revisiting topics considered in Chapters 0 and 3 from a more mathematical point of view. The expectation that the outcome of an addition always gives a greater value is deceptive, since adding zero to any number gives the same number in the end. The case of multiplication is more 'dramatic', since multiplication by 1 gives the same number and multiplication by 0 always results in 0. Moreover, with negative numbers or decimals, the expectation of a greater outcome when using addition or multiplication is even less appropriate. Multiplying 8 by ¼, for example, gives an answer smaller than 8.

Properties of operations on natural numbers

For the boffins among you: The number system of natural (whole non-negative) numbers, given by the set *N* of natural numbers together with a set of operations (addition and multiplication), of algorithms (subtraction and division), and relations (equality and order) on it, we call the '*N* system'. This may be represented by Figure 7.1.

Figure 7.1 A first landscape of natural numbers. Notice that we cannot represent all the natural numbers since there are infinitely many of them

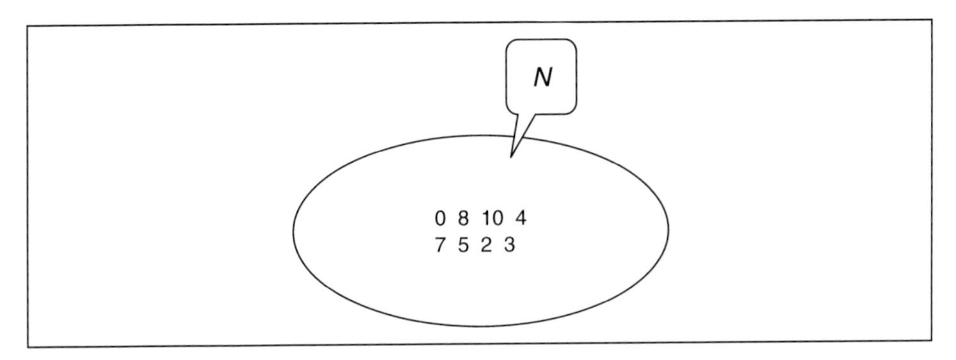

Addition and its properties

When we speak about properties of operations, we often think of the most familiar (and used) properties of addition and multiplication such as commutativity and associativity, even if we do not necessarily use these terms. The following examples allow us to focus on these properties and their use in everyday computations. The first is a double addition task inspired by Chapter 4.

To calculate $14 + 7 + 6$ we can work in a variety of ways. For example:

(a) going from left to right, $14 + 7 + 6 = (14 + 7) + 6 = 21 + 6 = 27$;
(b) going from right to left, $14 + 7 + 6 = 14 + (7 + 6) = 14 + 13 = 27$.

The equality $(14 + 7) + 6 = 14 + (7 + 6)$ is an example of the *associative property* of addition. The presence of this property allows both computations (a) and (b) above, giving the same result in each case. The use of brackets is unnecessary for addition and, indeed, we can add two, three, four or many more numbers without undue difficulty. (Compare this conclusion with the double division task of Chapter 4.) The presence of the associative property of addition also allows us to focus our attention on two numbers at a time before adding, for example, a third number to the result. For example, in (a) above we added 14 and 7 and then added 6 to the 21 which resulted from the first operation. Thus, we consider addition to be a binary operation.

Applying the laws of both commutativity and associativity we can also calculate $14 + 7 + 6$ as follows:

(c) $14 + (7 + 6) = 14 + (6 + 7) = (14 + 6) + 7 = 20 + 7 = 27$;
(d) $14 + (7 + 6) = (1 + 13) + (7 + 6) = 1 + (13 + (7 + 6)) = 1 + ((13 + 7) + 6) = 1 + (20 + 6) = (1 + 20) + 6 = 21 + 6 = 27$.

Examples (a), (b) and (d) use the associative property only, while the commutative property intervenes in (c).

We can employ the commutative and associative properties of addition in order to make computation easier, but they are inherent in the true nature of addition. Without these properties addition in columns is impossible.

To illustrate what we mean by the last assertion, think about $38 + 43$. This task could be performed in different ways (with different algorithms). We can obtain the sum by counting on, or considering the union of two (disjoint) sets having 38 elements and 43 elements, respectively, or performing addition in columns as in:

$$\begin{array}{r} 3\,8 \\ +\,4\,3 \end{array}$$

In this case, customarily, we first compute $(8 + 3)$ and then $(30 + 40)$, by a cultural convention; moreover, taking in account the amount carried, we calculate $(1 + 10) + 70 = 1 + (10 + 70) = 1 + 80 = 81$. Notice that we read words from left to right but, by another widespread cultural convention, column addition is performed from right to left. This is a consequence of the adopted place-value notation and the conventions (see Chapter 5).

In order to show the properties role in the column addition we can write meticulously, specifying the properties used: $38 + 43 = (30 + 8) + (40 + 3) =_{\text{(commutative)}} (8 + 30) + (3 + 40) =_{\text{(associative)}} 8 + (30 + (3 + 40)) =_{\text{(associative)}} 8 + ((30 + 3) + 40) =_{\text{(commutative)}} 8 + ((3 + 30) + 40) =_{\text{(associative)}} 8 + (3 + (30 + 40)) =_{\text{(associative)}} (8 + 3) + (30 + 40) = 11 + 70 = (1 + 10) + 70 =_{\text{(associative)}} 1 + (10 + 70) = 1 + 80 = 81$. Addition in columns is the most rapid and reliable procedure, but when we use place value all these applications of commutative and associative properties of addition are hidden.

Food for thought

Think about other examples in which these properties of addition might help a calculation.

Brackets tend to be useful for showing the order in which a calculation is done. As we shall see, however, they become essential when a number sentence includes more than one type of operation.

Before leaving addition, we can say that, on its own, it has the

- *closure property*, since for every pair of natural numbers a result can be obtained;
- *uniqueness property*, since for each pair of numbers their sum is uniquely given.

Figure 7.2 provides a diagrammatic summary of our discussion so far.

Figure 7.2 A landscape of *N* with addition. Notice that, as in Figure 7.1, we cannot represent the infinitely many addition calculations possible

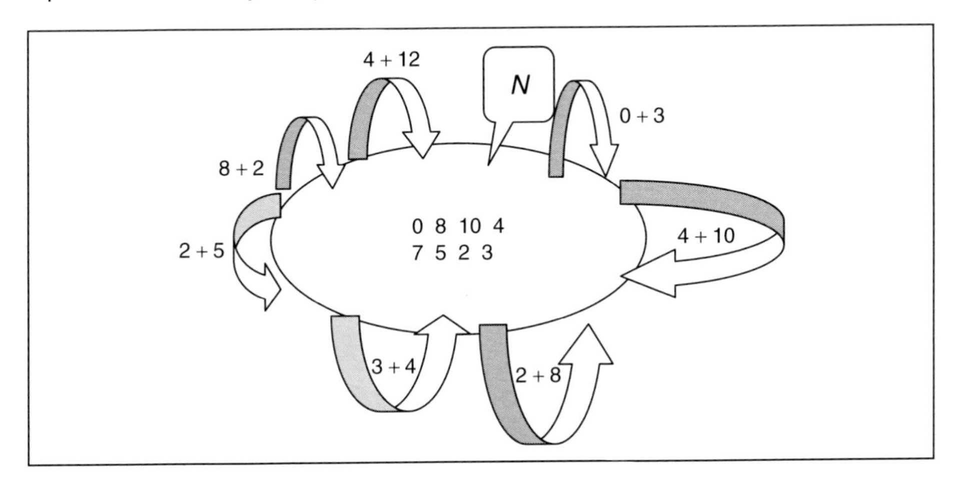

In order to appreciate these formal properties of addition, try replacing the addition signs in 14 + 7 + 6 with subtraction ones and note the strange and dreadful results!

(a′) $(14 - 7) - 6 = 7 - 6 = 1$;
(b′) $14 - (7 - 6) = 14 - 1 = 13$;
(c′) $14 - (7 - 6) = (?)\ 14 - (6 - 7) = 14 - (-1) = 14 + 1 = 15$;
(d′) $14 - (7 - 6) = (15 - 1) - (7 - 6) = (?)15 - (1 - (7 - 6)) = (?)\ 15 - ((1 - 7) - 6) = 15 - (- 6 - 6) = (?)\ (15 - 6) - 6 = 9 - 6 = 3$.

The results are all different! We had to put the equality sign followed by a question mark in some places since the two sides of such a relation are not equal. Note that some of these wrong computations cannot be performed within the *N* system, such as 6 – 7, 1 – 7 as, in both cases, negative numbers would result. Thus when subtraction is involved in the natural number system, we have to ascertain, firstly, whether the task is possible or not. Whenever the subtraction task is possible, then the result of the subtraction is unique.

Examples (a′) to (d′) confirm for us that subtraction is neither associative nor commutative. Moreover, certain subtraction calculations cannot be computed in the *N* system and those which can be computed give a unique result (difference).

We can affirm that in the *N* system subtraction does not have the *closure property*, but it satisfies the *uniqueness property*, that is, for every ordered pair of natural numbers for which it is possible to compute the difference. This is illustrated in Figure 7.3.

Figure 7.3 A landscape of N with addition and subtraction. Curly arrows indicate operations and algorithms that can be computed in N. The dotted arrow represents one of the many subtraction calculations which cannot be solved within the N system

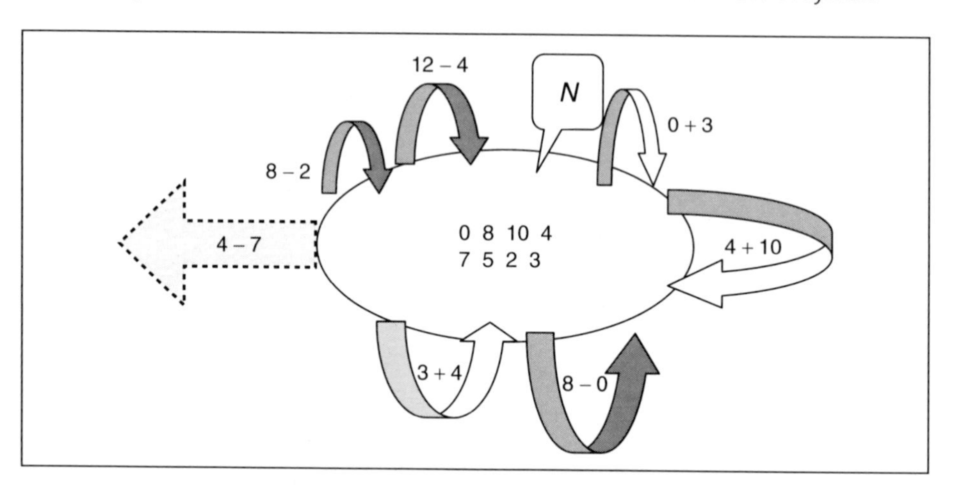

> **Challenge**
>
> Is zero a neutral element for addition and subtraction?
>
> What are the necessary properties for calculating subtraction in columns?

Multiplication and its properties

Calculate 36×8. This is a relatively simple task and we want to reflect upon the implicit and explicit assumptions we employ in order to perform it. We have discussed the properties of addition, and now we will take the addition case as a sort of outline for multiplication.

A first question is: are addition and multiplication similar with respect to the closure and uniqueness properties? Our first reaction is yes – in the case of 36×8 and also in all the cases obtained by arbitrarily changing 36 and 8 with two other natural numbers. We are asserting that multiplication, in the N system, has the closure property.

We are now in position to carry out the multiplication task. We can work in different ways. We could consider an array of 8 lines in each of which there are 36 things, or we could add $36 + 36 + 36 + 36 + 36 + 36 + 36 + 36$ – brackets are avoidable by the associative property of addition – or add 8 to itself 36 times, or we could multiply using the column method. Each computation gives 288 in the end.

We suspect that you will not be surprised by the coincidence of these results, obtained despite the use of different algorithms. This lack of surprise is due to the uniqueness property of multiplication. Figure 7.4 shows the new panorama of the N system.

Compute $4 \times 7 \times 6$. With these three numbers we can calculate the result in different ways:

Figure 7.4 A landscape of N with operations (addition, multiplication), and subtraction

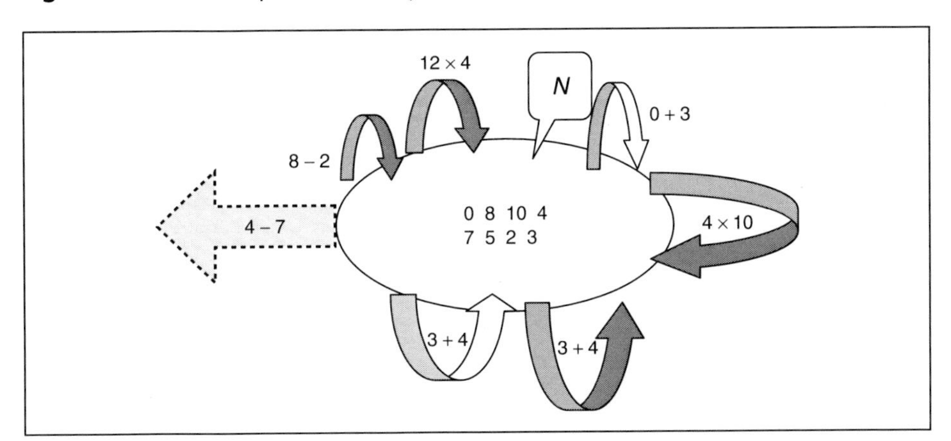

(a) going from left to right, $4 \times 7 \times 6 = (4 \times 7) \times 6 = 28 \times 6 = 168$;
(b) going from right to left, $4 \times 7 \times 6 = 4 \times (7 \times 6) = 4 \times 42 = 168$.

The equality $(4 \times 7) \times 6 = 4 \times (7 \times 6)$ is an example of the *associative property* of multiplication. The presence of this property allows both computations (a) and (b) above, obtaining the same product either way. In other words, we can choose whether or not to use brackets when multiplying three numbers. Indeed, we can multiply three, four, five or more numbers without problems (compare this conclusion with the double division task of Chapter 4). As with addition, the presence of the associative property of multiplication allows us to focus our attention on two numbers at a time.

Applying the laws of both commutativity and associativity, we can also calculate $4 \times 7 \times 6$ as follows:

(c) $4 \times (7 \times 6) = 4 \times (6 \times 7) = (4 \times 6) \times 7 = 24 \times 7 = 168$;
(d) $4 \times (7 \times 6) = (2 \times 2) \times (7 \times 6) = 2 \times (2 \times (7 \times 6)) = 2 \times ((2 \times 7) \times 6) = 2 \times (14 \times 6) = 2 \times 84 = 168$.

Examples (a), (b) and (d) are justified by the associative property only. The commutative property intervenes in (c).

Multiplication has the same properties as addition: closure, uniqueness, commutativity and associativity.

Consider how you would calculate 36×8 using the column method:

$$\begin{array}{r} 3\ 6 \\ \times\ 8 \end{array}$$

Firstly, you could calculate 6×8 and then 30×8 and add the two products. This procedure can be explained in this way: $36 \times 8 = (30 + 6) \times 8 = (6 + 30) \times 8 = (6 \times 8) + (30 \times 8) = 48 + 240 = 288$. Also multiplication using columns is the most rapid and reliable procedure. The equality $(6 + 30) \times 8 = (6 \times 8) + (30 \times 8)$ is possible because – as we will discuss below – of the *distributive property of multiplication over addition*.

Multiplication is distributive over addition. Without the distributive property of multiplication over addition we cannot compute multiplication in columns.

Challenge

Is multiplication distributive over subtraction?

Brackets are useful, but cumbersome. For the sake of conciseness, mathematicians adopt a convention: when addition and multiplication are involved, as in $14 + 7 \times 6$, first compute the multiplication and then the addition: $14 + 7 \times 6 = 14 + 42 = 56$. Notice that to calculate $(30 + 6) \times 8$ we cannot omit the brackets; if we do, given the above cultural convention, we would have to do the calculation thus: $30 + 6 \times 8 = 30 + (6 \times 8) = 30 + 48 = 78$.

We advise teachers to use brackets in the first approach to arithmetic. Our opinion is supported by the fact that the computations required of children are not long and the 'energy saving' obtained by omitting brackets is minimal. Think back to the Italian children in years 2–5 who were asked to complete the equality test in Chapter 1. When presented with a modified version with the same numbers but using brackets: the presence of brackets increased their rate of success.

A few words about division: it is well known that there are different interpretations and ways of calculating a division among natural numbers. In Chapter 4 different interpretations of division were presented, and it was demonstrated that division is not associative.

In the earlier years of schooling, calculations involving division are generally presented so that only natural numbers result (e.g. 8 divided by 2, 12 divided by 3), with the dividend greater than, and a multiple of, the divisor. These first examples can give a sort of imprinting and might be the origin of misconceptions. This warning is not intended to suggest that teachers must avoid such examples, but that it is possible to change perspective, asking pupils to reflect on the possibility that 8 or 12 are divisible also by other numbers (such as 1, 4), and the number 0 is divisible by 1, 4, 5, 7, and by 15 too, even though it is less than any of them. Another important fact is division by 1: it seems irrelevant since the quotient is equal to the dividend, but the role of 1 in division is central. This is discussed in the Appendix, as is the 'path' of division in primary school, from reception to year 6.

In order to enlarge the scope of division, in year 3 the so-called *Euclidean division* (division with remainders) is introduced. Remainders have a great importance in everyday life: we are accustomed to remainders when dividing by 7 (the days of the week) and by 12 or 24 (a timetable, the clock).

Challenge: A puzzling holiday

Our friend Ivan has a dacha (cottage) on a small island near Vladivostok harbour (Siberia). For summer holidays he starts from his working place, Moscow, using the Trans-Siberian railway to reach his dacha. The trip by train is 103 hours long. He knows that the ferry for the island leaves the harbour every two hours from 8 a.m. to 8 p.m. and it does not operate on Mondays. This year Ivan cannot leave Moscow until he has completed some important work. If he starts on Thursday 26 June 2008 at 2 p.m. does he have to book a room in a hotel in Vladivostok or could he sleep at home? If he has to make a reservation, how many nights should he book?

This strange problem can be solved with quotient and remainder: $14 + 103 = 117$, $117 \div 24 = 4$ (remainder 21). Therefore Ivan arrives Vladivostok at 21 hours (9 p.m., obtained by searching the remainder of the division $21 \div 12$), too late for the last ferry of the day, thus he needs a room in a hotel. He starts on Thursday, day 4 of the week, and the trip will take more than 4 days (the quotient of $117 \div 24$), therefore he will arrive on day $4 + 4 = 8$ of the week, but that means day 1 (the remainder of $8 \div 7$), or Monday, of the next week. Thus he has to book a room for one night only, since he could take the first ferry trip of Tuesday 1 July.

You can solve this problem by counting on (hours and days) but the use of remainders is nevertheless implicit in this manner.

The idea of considering both the quotient and remainder as a result of division increases the psychological 'cost', since we must leave the idea that an algorithm involving two (natural) numbers has one number as result. But, as you will see, it is a price worth paying. The idea comes from experience with addition and multiplication.

In fact, the result of a Euclidean division is a pair of numbers: the quotient and the remainder. Note that we cannot affirm that there are two results of the same division, but only that given two natural numbers (the dividend and the divisor) we get a unique pair (the quotient and the remainder) and this is *the* result of the division.

The necessary condition in order to obtain a unique pair (quotient, remainder) as the result of a Euclidean division is that the remainder must be greater than or equal to 0 and (strictly) less than the divisor.

It is simple to observe that the quotient and remainder of a Euclidean division can be determined for every pair of natural numbers (dividend, divisor), with the proviso that divisor is not 0 (see the Appendix).

Therefore, we cannot state that division has the closure property over every ordered pair of natural numbers, since 0 could be present as second component. In Italian schools – as in some other countries – division is mainly concentrated on the quotient, and in many cases the quotient is considered 'the result' of a division. If we cling to this interpretation, given an ordered pair of natural numbers (dividend, divisor), with the proviso that the divisor is different from 0, then we have the closure and uniqueness properties for the quotient. A summary of how division relates to the N system is given in Figure 7.5.

We conclude this section on operations with a specification about terminology. In the N system, addition and multiplication have the closure and uniqueness

Figure 7.5 A landscape of N with operations (addition, multiplication), and algorithms (subtraction, division). Curly arrows indicate operations and algorithms that can be computed in N. The new dotted arrow represents one of the infinite number of division calculations which cannot be solved within the N system

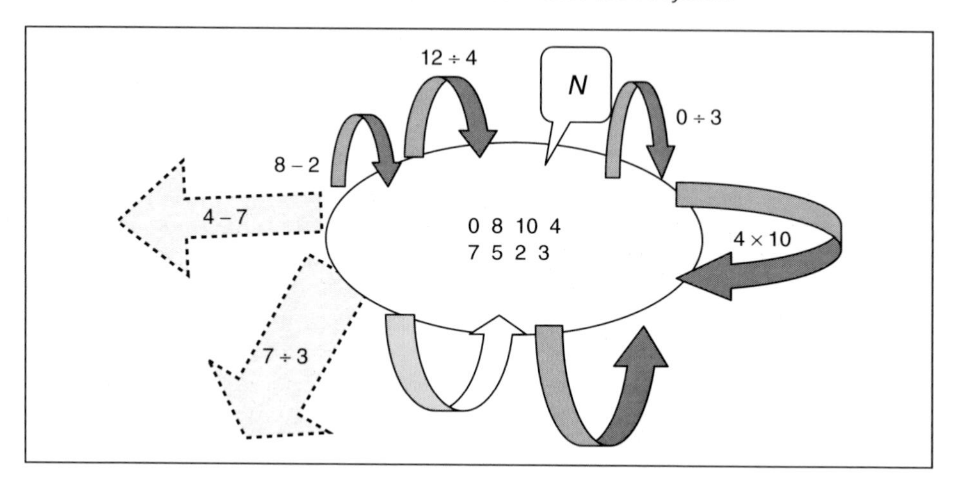

property; subtraction and division can be computed only if appropriate restrictive conditions are fulfilled. If subtraction and division can be computed, the uniqueness property also holds.

Addition, multiplication, subtraction and division act *functionally* on natural numbers, (by the uniqueness property), but we reserve the term *operation on natural numbers* only for addition and multiplication, since we can always perform them (closure property) without restrictive conditions on the terms.

This way of speaking could contrast with the widespread idea that we have four operations, but our proposal is more correct from the structural point of view. Addition and multiplication are not the only two operations on natural numbers, but to use the name 'operation' for subtraction and division is not completely correct.

Challenge

Find other examples of a procedure that can be applied to every pair of natural numbers and give a natural number as outcome.

Some solutions to this challenge are given in the Postscript below.

Properties of the equality relation on natural numbers

Pause for thought

When and why do we introduce the equality symbol in primary school?

When and why is the symbol '=' used in school? The symbol '=' is introduced and used early, in year 1, as a translation of the linguistic expression 'is equal to', 'the same number as' or 'as many as'. Usually we do not provide rich explanations for it, we use it when it is necessary or useful. We adopt it as it is a 'primitive term' or it has an intuitive meaning. From the theoretical point of view, '=' denotes a relation in N (or Z or ...). It is important to legalize its existence and its use: There is an equality relation in every number system. Chapter 1 is devoted mainly to equality, its uses and properties. We can now examine some different meanings for this symbol from a more mathematical point of view.

In the first years of primary school, the symbol '=' appears together with addition: if we need to express, for example, 'two plus one is equal to three', we write '2 + 1 = 3'. In this way, the '=' separates an operation (on the left) and its result (the sum, on the right). We go on with this particular meaning of the symbol, using it when we calculate subtractions, multiplications, divisions. It can create in the pupil's mind a particular idea of '=', that it is only a procedural sign, which marks the passage from an operation to its result. From this point of view, the expression '5 = 2 + 3' is intended differently from '2 + 3 = 5' and it can be considered 'strange'. Indeed, we usually employ the expression '5 = 2 + 3' with the aim of showing how it is possible to decompose 5 into the sum of a pair of numbers. Instinctively these points of view lead us to reject the symmetry of '='. But '2 + 3 = 5' can convey another idea: '5' and '2 + 3' are different ways of writing the same number. In this sense we can also write $2 + 3 = 1 + 4$ or $2 + 3 = 1 + 1 + 1 + 1 + 1$ or ... or $2 + 3 = 5 + 0$. We can search all the pairs of numbers whose addition makes 5: (0,5), (1,4), (2,3), (3,2), (4,1), (5,0). In Italy they are the 'friends of 5' and in the UK 'number bonds of 5'. From this point of view, we can write $1 + 4 = 2 + 3$. In this way, we do not distinguish the left from the right of the expression, using the *symmetrical property* of the relation '=' without difficulty. It is important to communicate that the relation '=' has the symmetrical property and, as a consequence, we can read an equality from left to right or vice versa. It is important to emphasise that '=' has the symmetrical property: if a is equal to b, then b is equal to a, since this property is structurally inherent in equality relations.

The reader could be flustered by this intrusion of letters instead of numbers both here and in earlier chapters, notably Chapters 2, 4 and 6. With this use of letters we gain in generality and clarity. In order to convince readers of our assumption, we will rewrite the same statement of the symmetrical property of equality relations, avoiding letters: '=' has the symmetrical property, meaning that given two numbers, if the first number is equal to the second, then the second number is equal to the first.

In this rhetoric form there could be another problem: children understand, intuitively, the phrase 'given two numbers' as 'given two *different* numbers', and therefore what follows could be meaningless. Moreover, the symmetrical property in itself does not require that there is a specific 'order' between letters a and b. The letters a and b, in the statement above, have the role of 'place holder', thus can be substituted appositely, even with the same number.

A final remark about the symmetrical property: this property is used and useful if we compare two different representations of the same number, and is completely useless if we consider the same representation, since in this case it could be considered a repetition of another property of the equality relation, the *reflexive property*. For example, if $3 + 5 = 2 \times 4$, then $2 \times 4 = 3 + 5$; in this form the symmetrical property guarantees that we can exchange one representation of 8 with another. But the statement 'if $8 = 8$, then $8 = 8$', which is a correct example of the symmetrical property of the equality relation, is sheer banality.

We now turn to the transitive property of equality: given three numbers, a, b and c, if a is equal to b and b is equal to c, then a is equal to c.

This could be an intuitive assertion, but it is also a basic and fundamental point about equality. For example, I have a lot of sticks and I classify them with reference to their length: if the length of the red stick is equal to the length of the grey stick and the length of the grey stick is equal to the length of black stick, I conclude that the red and black sticks have the same length. Given three playing cards ▣, ◙, ■, if I know that the number under ▣ is equal to the number under ◙ and I know also that the number under ◙ is equal to the number under ■, I can conclude that the number under ▣ is equal to the number under ■, even though I cannot see the numbers!

Euclid, following Aristotle's hypotheses, starts his *Elements* with some 'common notions'. The first one is: 'Things which equal the same thing also equal one another'. In this statement he summarises two properties together: symmetry and transitivity. We could translate this notion into symbols: if $a = c$ and $b = c$, then $a = b$. If $b = c$, indeed, $c = b$, for the symmetrical property of equality and if $a = c$ and $b = c$, then $a = c$ and $c = b$, therefore $a = b$ for the transitive property.

In Chapter 1 we refer to a test on equality given to pupils. In our experience, one of the most awkward exercises in this test was: $9 = \square$. Even for 15-year-old students it is very difficult to put only one number in the white square, all the more so because we must put in the square the same number! This is another characteristic of equality – the reflexive property of equality: every number is equal to itself. The sign '=' in this case acts like the pivot of a balance. What is on one side is balanced by what is on the other side.

Properties of order relations on natural numbers

Pause for thought

Does the knowledge and use of natural number sequences promote the idea of order in N or is it best to use the number line?

We can have many sequences of natural numbers. One such sequence is 0, 2, 1, 3, 4, 6, 5, 7, 8, 10, 9, 11, ... Another is 0, 1, 2, 3, 4, 5, 6, 7, 8, 9, 10, 11, ..., which, on account of its importance and to avoid being misled we can refer to as the *fundamental sequence*.

We start with an example: an Italian boy, who found mathematics very challenging, knew the fundamental sequence of natural numbers, but when looking for page 23 in a book, he would have to turn over each page from page 1 until he reached page 23. Evidently, for him, a knowledge of the fundamental sequence did not promote the concept of order between numbers. It is important to observe that if the natural numbers can be arranged in the fundamental sequence, it is because there is a *natural or fundamental* order relation on N, the relation 'less than' denoted by the symbol '<'. There is also the inverse relation 'greater than', denoted by the symbol '>'. These relations can be defined with respect to each other (*inter-defined*): if a first number is less than a second one, then the second number is greater than the first one and vice versa. If we stop and think about it, the comparison between numbers can be more important than the knowledge of the sequence. This, together with the equality relation, is shown in Figure 7.6.

Figure 7.6 A landscape of N with operations (addition, multiplication), algorithms (subtraction, division), and relations (equality, order)

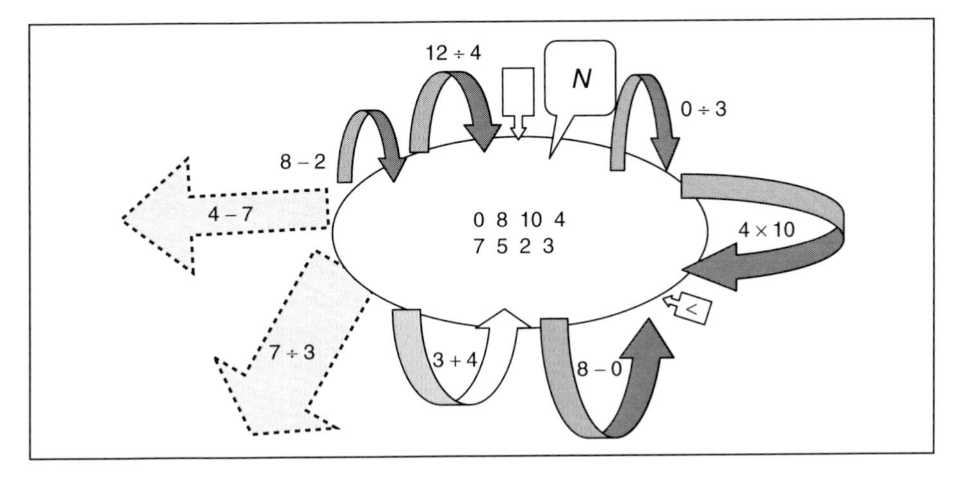

A fundamental property of order relations – the *comparison property* (or *trichotomy*) – can be stated as follows. Let a, b be two natural numbers. Then only one of these three different cases holds:

$$a < b \text{ or } a = b \text{ or } b < a$$

Furthermore, this *comparison property* can be reduced. If we know that a is different from b, then only one of the following two cases holds:

$$a < b \text{ or } b < a.$$

Let us use an example involving sticks. Imagine that some are 5 cm in length, some 10 cm and the rest 15 cm. We put the sticks into three different boxes A, B, C according to their lengths. As Piaget showed in a famous experiment, it is understandable that, when a stick in box A is shorter than a stick in box B and also a stick in box B is shorter than a stick in box C, then we know that a stick in box A is shorter than a stick in box C, without extracting a stick from each box. In other words, we have used the *transitive property* of the order relation '<': let a, b and c, be natural numbers, if a is less than b and b is less c, then a is less than c.

Challenge

Does the set N have a smallest element?

And does any non-empty subset A of N have a smallest element?

Is it possible to have two different smallest elements of a non-empty subset of N?

Look at an example. Consider the set A of all the natural numbers which are the odd multiples of 7, greater than 50 and less than 100. The set A is non-empty: since 77 is an odd multiple of 7, it is greater than 50 and less that 100. The number 63 also belongs to A, and 63 is a smallest element of A: looking at the 7 times table, we find that the multiples of 7 greater than 50 and less than 100 are 56, 63, 70, 77, 84, 91, 98; the odd ones are 63, 77, 91 only. Therefore 63 is a smallest element of A and it is unique, therefore we can state that 63 is *the* smallest element of A.

Our experience of natural number assures us that two different numbers cannot play the role of smallest element of a non-empty subset of N. Sometimes experience is deceptive: luckily not in this case. To convince the more sceptical readers, there is a simple proof of this in the Postscript at the end of this chapter.

The same argument applies to the case of 'largest'. Therefore, we conclude that the set N has the smallest element and each non-empty subset of N has a smallest element unique to that subset. Obviously 0 is the smallest natural number.

To search for the largest number, we can continue to count ... never stopping, thus the largest natural number does not exist.

A non-empty subset of N does not always have a maximum. An example is given by the subset of even numbers.

A more complex notion is that of a non-empty superior limited subset. Let A be a non-empty subset of a set of natural numbers B. The set A is *superior limited in B* if there exists a natural number b, belonging to B, such that for every a belonging to A, $a \leq b$.

For example, consider the set B of odd natural numbers and its subset A of the odd natural numbers which are multiples of 3 and less than 24. Then, 3, 9, 15, 21 are (all) the elements of A. We can affirm that A is non-empty and superior limited in B, since 35 belongs to B (but not to A) and 35 is such that for every a belonging to A, $a < 35$, that is, $3 < 35$, $9 < 35$ ... , $21 < 35$. We can also state that 21 is the greatest element of A, the maximum of A, since 21 belongs to A and it is greater than each other element belonging to A.

What happens for these sets A and B occurs in general: each non-empty subset A of natural numbers superior limited in N has a maximum. This property is not usually considered in primary schools, but it becomes highly relevant in secondary schools, since it fails with rational numbers.

A more familiar property of order on natural numbers is the following. Let p be a natural number. Then $p+1$ is the next number. Does there exist a natural number s such that $p < s$ and $s < p+1$? More explicitly, can we find a natural number a such that $49 < a < 50$, or $1000 < a < 1001$? The answer is negative because in N the order $<$ is discrete, thus N with this order is called a discrete set.

The relations <, >, ≥, ≤ are the most important order relations for *N*, but not the only ones. When we enlarge the number system of natural numbers to other richer number systems, we introduce many extensions of these order relations.

Conclusions about the ℕ system

We have described the *N* system in some detail, presenting the main formal properties of operations (addition and multiplication), algorithms (subtraction and division) and relations (equality and order). The formal properties are important since they make mathematics possible. We can consider these formal aspects from two points of view:

- they are a sort of frill, owing to the pique of mathematicians used to splitting hairs, but (school) mathematics is something else;
- they are the true essence of mathematics since they offer an appropriate answer to the difficult question: what is mathematics? (MacLane, 1986).

Your efforts in following us in this long description of what the *N* system is have not been in vain since this system is a sort of outline for the other number systems.

Other number systems

Note that our presentation of the *N* system diverges from standard practice in using the term 'algorithm' for subtraction and division, rather than 'operation', since we wish to stress the fact that these procedures have exceptions, that is, they do not have the property of closure since they cannot be applied to every ordered pair of natural numbers to get a natural number as a result. These difficulties in applying subtraction and division (with remainder 0) are the starting point for the construction of other number systems.

The ℤ system

Consider the game shown in Figure 7.7, which was presented in the *Eastern Daily Press* newspaper in October 2007. The title of the game could be seen as misleading as it includes subtraction and multiplication as well as addition. Moreover, the text mentions division, but division is not required in the game.

From a mathematical point of view, the game (without accompanying text) asks for the solution of an algebraic system with six equations and nine unknowns. Under these conditions there are infinitely many possible solutions, two of which are given in Figure 7.8.

The accompanying text suggests that we must stick to natural numbers, bearing in mind the rules for each operation. But if you look at the last column closely you will see that it requires three numbers – call them a, b, and c – such that $a - b + c = 1$. This statement could be ambiguous, since the puzzle text does

Figure 7.7 The newspaper game

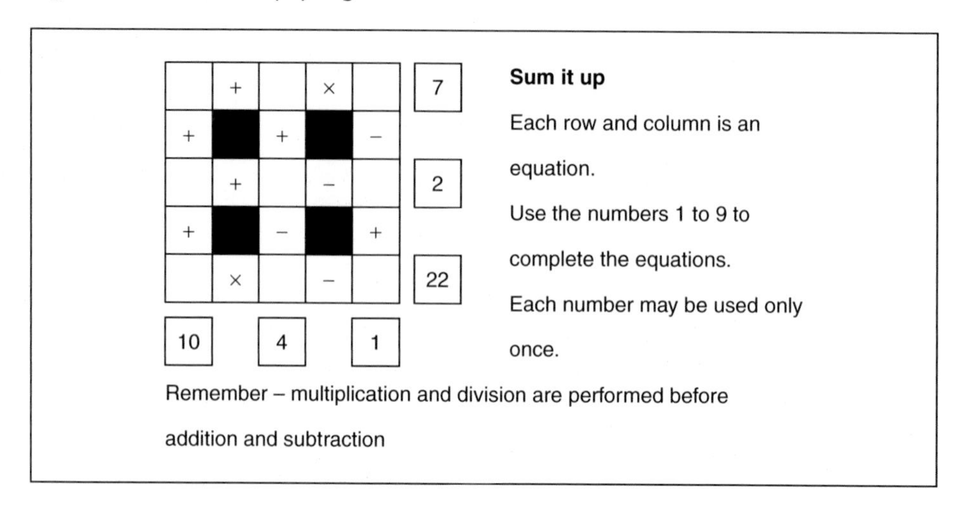

Figure 7.8 Possible solutions of the game (disregarding accompaying text conditions)

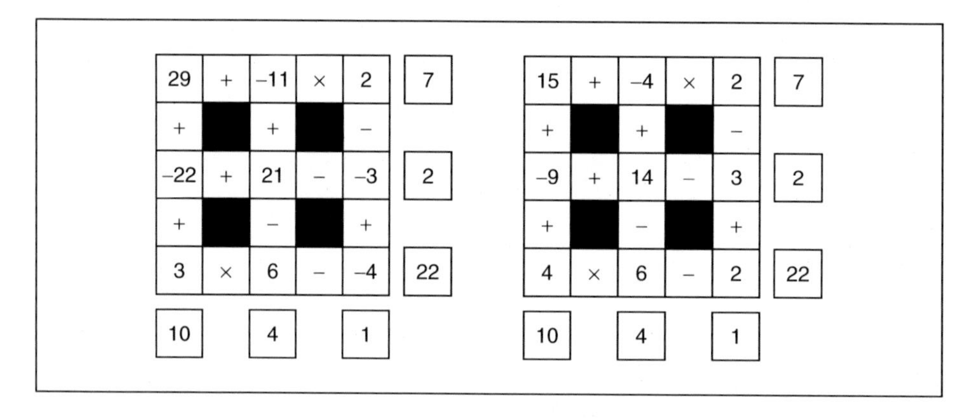

not specify how addition and subtraction must behave, so we can ask ourselves if we have to do subtraction first and then addition, or addition first and subtraction later. Using brackets, the two ways of computating can be read as (a – b) + c = 1 (the first alternative) or a – (b + c) = 1 (the second alternative). From the given rule that each number from 1 to 9 must be used only once in the scheme, adopting the first reading of the requirement, $(a - b) + c = 1$, we have the sum of two natural numbers is 1 if and only if one of them is 1 and the other one is 0. But under the text conditions c cannot be 0, thence $c = 1$ and $a - b = 0$, but this equality is impossible, since it implies $a = b$, violating the 'only once' rule. Even if we read $a - (b + c) = 1$ the task is impossible since in the upper right corner box of the first row of the game should be filled with a number less than 7. We cannot use 6, otherwise we obtain 7 only as $1 + 1 \times 6$, violating the 'only once' rule. Therefore, we can fill the upper right-hand box of the scheme with 5, 4, 3, 2 or 1 only. But $a = 1$ implies $b + c = 0$ which is impossible since b, c must be two numbers greater that 0; $a = 2$ implies $b + c = 1$, which is impossible, since 0 is not allowed and from $b + c = 1$, we get

$b = 1$ and $c = 0$ or $b = 0$ and $c = 1$; $a = 3$ implies $b + c = 2$, but for 'only once' rule we cannot have $b = 1 = c$, and 0 is not allowed. Two cases $a = 4$ or $a = 5$ remain in order to have 7 as the result of first row, we must consider $7 = 3 + 1 \times 4$. For the third column, from $a = 4$, we have $b + c = 3$, and this is possible only if 1 is involved, violating the 'only once' rule. If $a = 5$, then the first row is given by $2 + 1 \times 5$, and $b + c = 4$. Even in this case the task is impossible to solve.

In our opinion, to solve the puzzle the last column must be read as $(a - b) + c = 1$, even if this is impossible using only natural numbers, with $a = 2$, $b = 9$ and $c = 8$. In this case $a - b = -7$ which is not a natural number!

The complete solution (devised by the newspaper) is unique and is shown in Figure 7.9. We must extend the N system in order to solve the game. But we have rather more serious motives for doing so.

Figure 7.9 The (newspaper) solution of the game

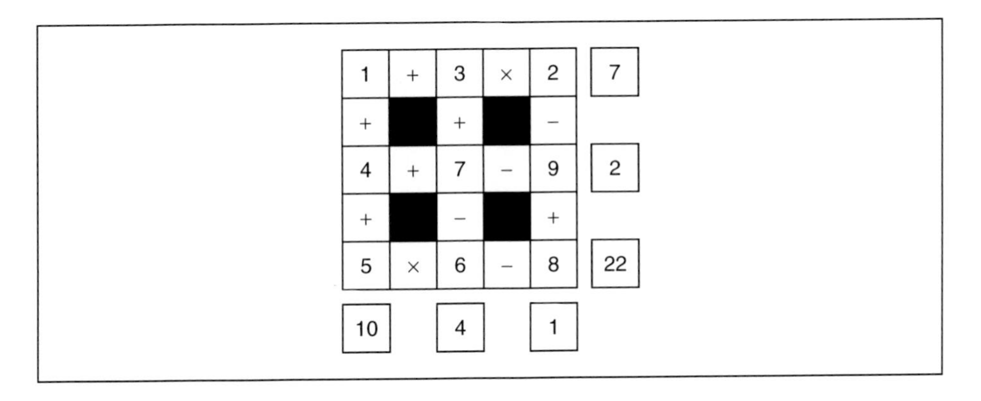

Subtraction on natural numbers does not have the closure property, since we can only apply subtraction to suitable ordered pairs of natural numbers: we can compute the difference $9 - 3 = 6$, but we cannot calculate (with natural numbers) $3 - 9$. Moreover, subtraction is neither commutative nor associative.

The condition of applicability for subtraction is that the first number (the minuend) must be greater than or equal to the second number (the subtrahend). But there are everyday circumstances where we need to compute differences such as $3 - 9$: when the temperature is 3°C and, due to a cold wind, the temperature drops 9 degrees. In bank accounts numbers can 'change' colour, when outgoings exceed deposits.

The convention mathematicians follow in such circumstances (i.e. when the minuend (say, 3) is smaller than the subtrahend (say, 9), is to compute the difference between the smaller and the larger and put a minus sign in front of the resulting number (here –6). Such a strategy reminds you of the numbers involved, and that they are the result of a subtraction process which you cannot perform on natural numbers. This use of symbols is firmly established in mathematics, but as some people find the use of '–' confusing when it is an integral part of the answer, some authors write '→3' instead of '+3' and '←5' instead of '–5'. This relates to the ideography of the number line when, for example, you consider negative numbers. Minus three, minus four, and so on, are considered to be representations of a new kind of numbers, the negative numbers. If you want, we can continue our number line to the left of zero, making a sort of symmetry of the already familiar number line used to represent natural numbers.

This way of introducing negative numbers by symmetry is cosy, it is a very useful way to define the order naturally, but it does not facilitate the arithmetical properties of these numbers, mainly when multiplication and division are involved. From this latter point of view the best approach is the one taken by the problem of difference between two natural numbers.

Figure 7.10 A diagram illustrating the building up of the *N* and *Z* number systems

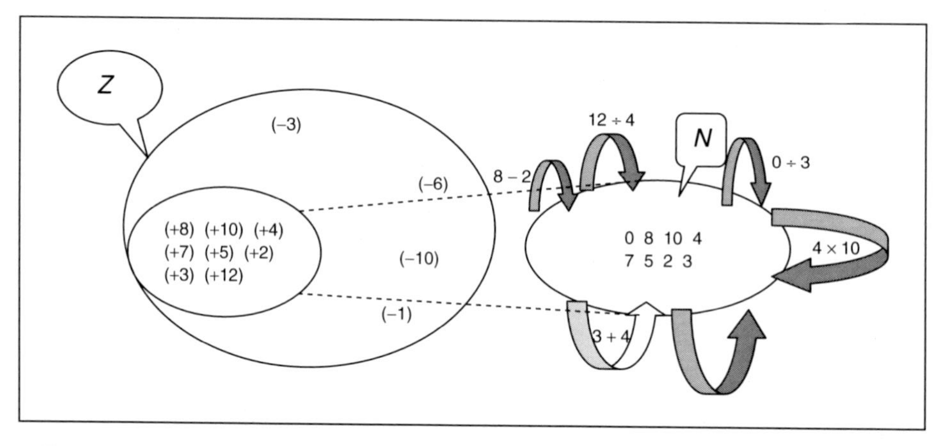

Figure 7.10 can be considered a sort of strange geographical map of the *N* system and *Z* system, with their connections. This can then be expanded into Figure 7.11, which provides more detail.

Figure 7.11 *N* and *Z*, their properties and the connections between them

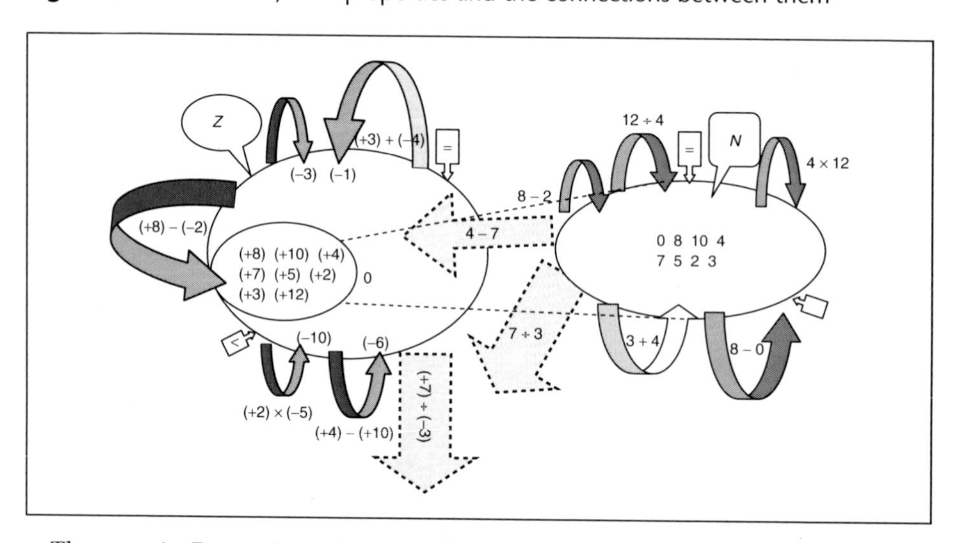

The map in Figure 7.11 shows the 'towns' *N* and *Z* with their walls, and the balloons are the road signs with the town name. It shows the inhabitants of the two towns (the numbers). The curly arrows are (examples of) activities performed by the inhabitants and having a result in the same town. Each inhabitant of $\mathbb{N}$ has a twin living in $\mathbb{Z}$ – for example, the twin of 8 living in $\mathbb{N}$ is +8 living in $\mathbb{Z}$. The connection between the two towns on the map is a dotted arrow. In fact there is an infinite number of sets of such bridges connecting $\mathbb{N}$ and $\mathbb{Z}$ and we show only one of them, 4 – 7, otherwise the map would become indecipherable. There is a dotted arrow going outside of *N* but not ending in Z, and a dotted arrow from $\mathbb{Z}$: they stand for ways to another 'town' (off the map). We represent the relations between the inhabitants of the same town with ⇨ arrows. These arrows are labelled with the names of equality and order relations.

The search for a realm in which subtraction has the closure property is successful: in the *Z* system every subtraction calculation can be performed without conditions on the terms.

The $\mathbb{Q}$ system

We begin with a game (Figure 7.12) similar to the newspaper one of Figure 7.7. Figure 7.11 leaves open the question of 7÷3 (and (+7) ÷ (–3)), represented by two dotted arrows. We know that following the Euclidean division algorithm 7 ÷ 3 gives the quotient 2 and the remainder 1 (and the Euclidean division in *Z* gives (+7) ÷ (–3) = (–3) with remainder (+2)). But now we want a division avoiding the oddity of the result given by a pair of numbers. We also desire a division with remainder 0, whatever the dividend and divisor (different from 0), and therefore we only need to consider the quotient.

Figure 7.12 Carlo's game

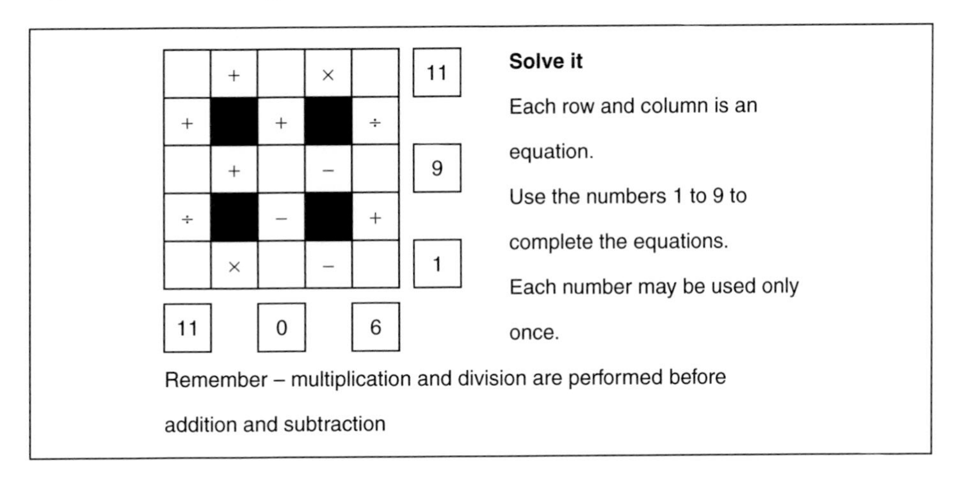

For this problem we have a simple solution: to use the problem as solution of itself. This is the solution proposed in 1202 by Leonardo Pisano, known also as Fibonacci, in his *Liber Abbaci*. Following him, we may denote the result of the division 7÷3 simply as 7/3, another ancient way of representing division, without calculating it. We do not explain what this new mathematical object must be and describe its nature.

We can organise operations using symbols of this sort in order to treat them as numbers. As you probably know, these symbols are called *fractions*, and the names related with them are *numerator* and *denominator*. In our example of 7/3, 7 is the *numerator*, and 3 is the *denominator*.

Most children are aware of fractions from an early age when, for example, telling the time: half past three, a quarter to five, and so on. We become familiar with $\frac{1}{2}$ and $\frac{1}{4}$ when we are introduced to the geometrical shape of the clock face and its symmetrical properties. Only later do we learn that these symbols are not to be considered 'as a whole', but the numbers in them have their specific names: *numerator* and *denominator*. These names come from Latin: numerator (*giving the number*) and denominator (*giving the name*). Thus $\frac{7}{3}$ is read as 'seven thirds', 7 is the numerator, and the expression of quantity relating to the choice of magnitudes, in this case expressed by 3, is the denominator. With the exception of twos ('halves') and fours ('quarters'), the thing giving the name – the corresponding ordinal adjective – is easily matched to the symbol it represents. Let us take two more fractions, and compare them to $\frac{7}{3}$. Given $\frac{7}{4}$ and $\frac{5}{3}$, we read the first new fraction as 'seven quarters' and the second as 'five thirds'. The fractions $\frac{7}{3}$ and $\frac{7}{4}$ tell us that we are considering the same quantity (7), but related to different magnitudes (identified by denominators 3 and 4, respectively). If we compare these, we find that $\frac{7}{4}$ is less than $\frac{7}{3}$, and this fact might result in a misconception on the part of a child. In the case of $\frac{7}{3}$ and $\frac{5}{3}$ we have different quantities (given by numerators 7 and 5) of the same magnitude, expressed by 3. In this case comparing the fractions is simply a matter of comparing the quantities, hence $\frac{5}{3}$ is less than $\frac{7}{3}$.

We get equivalent fractions $\frac{7}{3}$ and $\frac{21}{9}$ by multiplying both numerator and denominator of the first by the same factor (3). We call this equivalence the invariance property of division.

Notice that the invariance property for Euclidean division in N holds and does not hold at the same time: if the problem is to simplify the computation, we can consider $37 \div 5$ instead of $259 \div 35$, since $259 = 37 \times 7$ and $35 = 5 \times 7$. Both have the quotient 7 (invariance property), but the remainder of the first division is 2 and the remainder of the second is 14. Therefore the invariance property holds for the quotient, but does not hold for remainder, in the case of a remainder different from 0. But with fractions the remainder is always $0 : \frac{7}{3} \div \frac{2}{5} = \frac{35}{6}$, hence the invariance property holds.

In the same way rational numbers include all the possible equivalent fractions under the invariance property.

For other purposes we can follow another, simpler, way sacrificing the 'accuracy' for practical aims. Consider multiplication in which the multiplier is 10, or 100, or 1000, the products are easy to find: $4 \times 1000 = 4000$; $32 \times 100 = 3200$; $103 \times 10 = 1030$ (and not 130). The rule is simple: when the multiplier is 'one followed by zeros' the product has as many zeros as the multiplier has, assuming there are no zeros in the multiplicand.

This rule is customary, but the last case and the example of $2010 \times 10 = 20\,100$ (and not 2100) shows that we cannot always apply it without making a mistake. We can say (correctly) that when the multiplier is of the form 1 followed by zeros, we obtain the product by extending the writing of the multiplicand with as many zeros as there are in the multiplier.

We can state a similar rule with division, if the divisor is 10, 100, 1000, … . Instead of adding we can think in terms of deleting the zeros, as in 3200 ÷ 10 = 320, 4000 ÷ 1000 = 4 and so on.

But what can we do in the case 3200 ÷ 1000? This is an impossible division in *N* and in *Z*, and we can indicate the result as $\frac{3200}{1000}$ or equivalently $\frac{32}{10}$. It is customary to write the second fraction as 3.2. The same goes for 28 ÷ 10 = 2.8. In this case we use the so-called *decimal point* and the numbers associated with it are called *decimal numbers*. Decimal numbers are formally introduced in year 4, but children are usually aware of them long before, through the use of money.

Decimal numbers are difficult to treat with operations and order relations. Moreover, if a CD costs £9.99 we can consider this cost as £10, introducing a sort of equality by indifference, based on approximation.

Note that we can consider a decimal number as the quotient of 7 ÷ 3, but, if we do, we obtain 7 ÷ 3 = 2.33333… and so on and so on endlessly. However, if we stop the division, say, at the third decimal digit, we can assume 2.333 as the result of division 7 ÷ 3, with an approximation. This example shows that in many cases fractions are more precise than decimal numbers and often these latter can only be considered approximate values. For many practical tasks it is enough to stop at the second decimal digit. Also a decimal number is sometimes the solution to an impossible division problem on natural numbers, therefore it is not a natural number!

Therefore unsolvable division problems on natural numbers give rise to another numerical system. We denote it by the letter *Q* (from quotient!) and its elements are called *rational numbers*.

This presentation is too rough. The numerical system of rational numbers is not built starting from impossible division on natural numbers, but is the result of impossible division problems on relative integer numbers.

The construction of Q, the set of rational numbers can be done in different ways. A drawing can help us (see Figure 7.13). This also includes the so-called *absolute rational* number system, denoted Qa. In order to express the reciprocal relations among *N*, *Z*, Q_a and *Q* more clearly, we have omitted the elliptical shapes with which sets are customarily represented.

The arrows in Figure 7.13 show the connections among these number systems and also indicate how each number system fits into a richer one. The dotted arrow connecting *N* and *Q* is the composition of two solid arrows – down then left, or left then down. The 'surprise' is that the two paths have the same effect.

Figure 7.13 is not a complete representation of the number system but it covers the fundamental requirements for primary schooling, making it a good place for us to stop.

Summary tables

Now we are in the presence of new number systems and we need to explore them with the instruments mathematics offers us, comparing them with our

Figure 7.13 *The number systems N, Z and Q, including the absolute rational numbers Q_a*

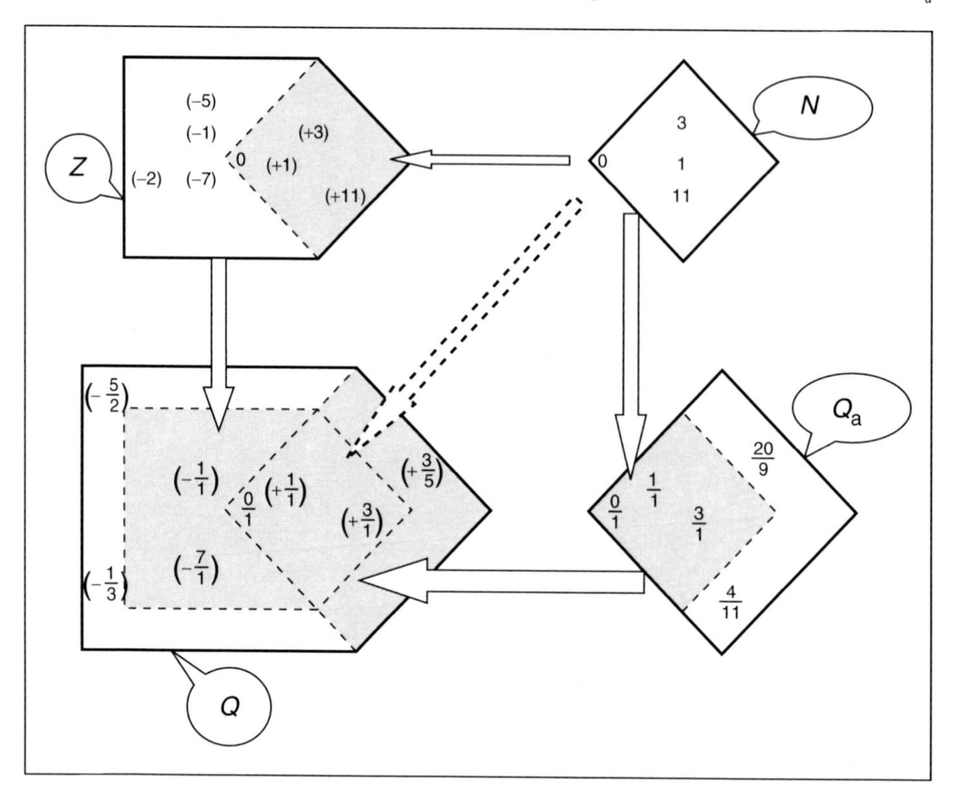

knowledge about the *N* system. Tables 7.1 and 7.2 might act as a sort of guide into these unknown countries.

The tables summarise the principal aspects of the operations and relations in the three number systems *N*, *Z* and *Q*. These tables might be useful for teachers wanting a quick reference point or wishing to deepen their mathematical knowledge. We have included some theoretical definitions which are useful for understanding the properties quoted in the tables. Some of them we have already discussed but, for the sake of completeness, it is useful to have these definitions collected together.

To understand aspects of the tables, let A be a non-empty set of numbers included in B. The set A is superior limited in B if there exists a number b, belonging to B, such that for every a belonging to A, $a \leq b$. The set A is inferior limited in B if there exists a number b, belonging to B, such that for every a belonging to A, $b \leq a$. When B is *N* or Z or Q we skip the specification 'in N' or 'in Z' or 'in Q'.

Table 7.1 presents a non-exhaustive list of the property of equality and orders in *N*, *Z* and in *Q*. Table 7.2 treats algorithm and operations. Remember that *operation*, in the terminology adopted, means *an algorithm with closure and uniqueness properties*.

Table 7.1 *Equality and order relations in* ℕ, ℤ *and* ℚ

Relations	Property	Resides in		
		ℕ	ℤ	ℚ
=	Existence: *There is an equality relation on the set of numbers. In symbols, =*	Yes	Yes	Yes
=	Reflexive: *each number is equal to itself.*	Yes	Yes	Yes
=	Symmetrical: *for every a, b, if a = b, then b = a.*	Yes	Yes	Yes
=	Transitive: *for every a, b, c, if a = b and b = c, then a = c.*	Yes	Yes	Yes
=	Substitutive: *for every a, b, if a = b, everything we do with a, we can do with b, obtaining the same things.*	Yes	Yes	Yes
< (>)	Existence: *there is a (strict) order relation – 'less than' ('greater than'), in symbols < (>).*	Yes	Yes	Yes
< (>)	Inter-definition: *for every a, b, if a < b, then b>a and conversely*	Yes	Yes	Yes
< (>)	Transitive: *for every a, b, c, if a < b and b < c, then a < c (if a > b and b > c, then a > c).*	Yes	Yes	Yes
< (>)	Discrete: *for every a there is no number b such that a < b < a +* 1.	Yes	Yes	No
< (>)	Dense: *for every a, b such that a < b, there is a number c such that a < c < b.*	No	No	Yes
< (>)	Total order: *for every a, b, if a ≠ b, then a < b or b < a, but not both.*	Yes	Yes	Yes
< (>)	Minimum existence: 0 *is the least number.*	Yes	No	No
< (>)	Maximum existence: *there is the greater number.*	No	No	No
< (>)	Minimum existence for non-empty subsets: *Every non-empty subset has a minimum.*	Yes	No	No
< (>)	Maximum existence for non-empty superior limited subsets: *for each non-empty superior limited subset A, there is an element of A which is the maximum of A.*	Yes	Yes	No
< (>)	Minimum existence for non-empty inferior limited subsets: *for each non-empty inferior limited subset A, there is an element of A which is the minimum of A.*	Yes	Yes	No
≤ (≥)	Existence: *there is a (non-strict) order relation 'less than or equal to' ('greater than or equal to'). In symbols ≤ (≥).*	Yes	Yes	Yes

Table 7.1 (Continued)

Relations	Property	Resides in		
		N	Z	Q
$\leq (\geq)$	Inter-definition: *for every, a,b, if* $a \leq b$, *then* $b \geq a$ *and conversely.*	Yes	Yes	Yes
$\leq (\geq)$	Reflexive: *each number is less than or equal to itself.*	Yes	Yes	Yes
$\leq (\geq)$	Antisymmetrical: *for every a, b, if* $a \leq b$ and $b \leq a$, *then* $a = b$ (*if* $a \geq b$ and $b \geq a$, *then* $a = b$).	Yes	Yes	Yes
$\leq (\geq)$	Transitive: *for every a, b, c, if* $a \leq b$ *and* $b \leq c$, *then* $a \leq c$ ($a \geq b$ *and* $b \geq c$, *then* $a \geq c$).	Yes	Yes	Yes
$\leq (\geq)$	Discrete: *for every a if b is such that* $a \leq b \leq a+1$, *then* $b = a$ *or* $b = a+1$, *but not both.*	Yes	Yes	No
$\leq (\geq)$	Total order: *for every a, b,* $a \leq b$ *or* $b \leq a$.	Yes	Yes	Yes
$\leq (\geq)$	Minimum existence: 0 *is less than or equal to any number.*	Yes	No	No
$\leq (\geq)$	Maximum existence: *there is a number such that it is greater than or equal to any number.*	No	No	No
$\leq (\geq)$	Minimum existence for non-empty subsets: *Every non-empty subset has a minimum.*	Yes	No	No
$\leq (\geq)$	Maximum existence for non-empty superior limited subsets: *for each non-empty superior limited subset A, there is an element of A which is the maximum of A.*	Yes	Yes	No
$\leq (\geq)$	Minimum existence for non-empty inferior limited subsets: *for each non-empty inferior limited subset A, there is an element of A which is the minimum of A.*	Yes	Yes	No

The huge list in Table 7.2 is what we regard as fundamental for primary school (teachers') knowledge, but it is far from exhaustive. Number theory is one of the most fascinating fields of mathematical research.

Table 7.2 *Algorithms in* $\mathbb{N}$, $\mathbb{Z}$ *and* $\mathbb{Q}$

Algorithm	**Property**	*N*	*Z*	*Q*
subsequent	Existence: *there is a (unary) subsequent algorithm on the set of numbers.*	Yes	Yes	No
subsequent	Closure: *for every number the subsequent algorithm gives a number in the end.*	Yes	Yes	No
subsequent	Uniqueness: *for every number the subsequent algorithm gives a unique number.*	Yes	Yes	No
subsequent	0: *the number* 0 *is the subsequent of a number.*	No	Yes	No
precedent	Existence: *there is a (unary) precedent algorithm on the set of numbers.* (Every natural number different from 0 has a precedent)	No	Yes	No
precedent	Closure: *for every number the precedent algorithm gives a number in the end.*	No	Yes	No
precedent	Uniqueness: *for every number the precedent algorithm gives a unique number.*	Yes	Yes	No
+	Existence: *there is a (binary) addition algorithm on the set of numbers. In symbols* +	Yes	Yes	Yes
+	Closure: *for every couple of numbers, the addition algorithm gives a number in the end.*	Yes	Yes	Yes
+	Uniqueness: *for every couple of numbers, the addition algorithm gives a unique number called the sum of the two numbers.*	Yes	Yes	Yes
+	Commutative: *for every* a, b, $a + b = b + a$.	Yes	Yes	Yes
+	Associative: *for every* a, b, c, $(a + b) + c = a + (b + c)$.	Yes	Yes	Yes
+	Existence of neutral element: 0 *is a number and for every* a, $a + 0 = a$.	Yes	Yes	Yes
+	Reduction: *given* a,b,c *numbers, if* $a + c = b + c$, $a = b$.	Yes	Yes	Yes
+	Existence of the opposite: *for every* a *there is a* b *such that* $a + b = 0$. The natural number 0 is an exception since $0 + 0 = 0$, but no number different from 0 has an opposite.	No	Yes	Yes

(Continued)

Table 7.2 (Continued)

Algorithm	Property	N	Z	Q
+	Increase: *for every a, b,* $a + b \geq a$ and $a + b \geq b$.	Yes	No	No
+	Unitary sum: *if* $a + b = 1$, *then either* $a = 1$ *and* $b = 0$ *or* $a = 0$ *and* $b = 1$, *but not both.*	Yes	No	No
×	Existence: *there is a (binary) multiplication algorithm on the set of numbers. In symbols* ×.	Yes	Yes	Yes
×	Closure: *for every pair of numbers, the multiplication algorithm gives a number.*	Yes	Yes	Yes
×	Uniqueness: *for every pair of numbers, the multiplication algorithm gives a unique number called the product of the two numbers.*	Yes	Yes	Yes
×	Commutative: *for every a, b,* $a \times b = b \times a$.	Yes	Yes	Yes
×	Associative: *for every a, b, c,* $(a \times b) \times c = a \times (b \times c)$.	Yes	Yes	Yes
×	Existence of a neutral element: 1 *is a number and for every a,* $a \times 1 = a$. More precisely, the neutral element of multiplication in Z is +1 and in Q is $\left(+\frac{1}{1}\right)$	Yes	Yes	Yes
×	Existence of an absorbent element: 0 *is a number and for every a,* $a \times 0 = 0$.	Yes	Yes	Yes
×	Reduction: *for every a,b,c, different from* 0, *if* $a \times c = b \times c$, *then* $a = b$.	Yes	Yes	Yes
×	Existence of the inverse: *for every number a there is a number b such that* $a \times b = 1$. The natural number 1 is an exception since $1 \times 1 = 1$, no other natural number different from 1 has an inverse. For relative integer numbers (+1) and (–1) are exceptions since $(+1) \times (+1) = (+1)$, and $(-1) \times (-1) = (+1)$; no other integer number different from these two has an inverse; a rational number a has an inverse if and only if $a \neq 0$.	No	No	No
×	Existence of the inverse for non-zero numbers: *for every non-zero number a there is a number b such that* $a \times b = 1$.	No	No	Yes
×	Annulment: *the product* $a \times b = 0$ *if and only if* $a = 0$ *or* $b = 0$ *or both.*	Yes	Yes	Yes

Table 7.2 (Continued)

Algorithm	Property	N	Z	Q
×	Unitary products: *if* $a \times b = 1$, *then* $a = 1$ *and* $b = 1$. For relative integer numbers, if $a \times b = (+1)$, then $a = +1$ and $b = +1$ or $a = -1$ and $b = -1$, but not both.	Yes	No	No
×	Increase: *for every* a,b, $a \times b \geq a$ *and* $a \times b \geq b$.	No	No	No
×	Increase for non-zero numbers: *let* a,b *be different from* 0, *then* $a \times b \geq a$ *and* $a \times b \geq b$.	Yes	No	No
+, ×	Distributive: *for every* a,b,c, $a \times (b + c) = (a \times b) + (a \times c)$.	Yes	Yes	Yes
–	Existence: *there is a (binary) subtraction algorithm on the set of numbers. In symbols* –. Some subtractions can be performed on natural numbers. The answer 'Yes' in column N and in the following rows refers to properties of the possible subtractions.	No	Yes	Yes
–	Closure: *for every pair of numbers, the subtraction algorithm gives a number.*	No	Yes	Yes
–	Uniqueness: *for every pair of numbers, the subtraction algorithm gives a unique number called the difference of the two numbers.*	Yes	Yes	Yes
–	Commutative: *For every* a, b, *then* $a - b = b - a$.	No	No	No
–	Associative: *For every* a, b, c, $(a - b) - c = a - (b - c)$.	No	No	No
–	Existence of right neutral element: 0 *is a number and for every* a, $a - 0 = a$.	Yes	Yes	Yes
–	Existence of left neutral element: 0 *is a number and for every* a, $0 - a=a$.	No	No	No
–	Reduction: *For every* a, b, c, *if* $a - c = b - c$, *then* $a = b$; *and if* $c - a = c - b$, *then* $a = b$	Yes	Yes	Yes
–	Decrease: $a - b \leq a$.	Yes	No	No
–, ×	Distributive: *for every* a,b,c, $a \times (b - c) = (a \times b) - (a \times c)$.	Yes	Yes	Yes
÷	Existence avoiding zero divisors: *there is a (binary) division algorithm on the set of numbers.* Some divisions with non-zero divisors can be performed on natural numbers, some not. The same happens on relative integer numbers.	No	No	Yes

Table 7.2 (Continued)

Algorithm	Property	*N*	*Z*	*Q*
÷	Existence of Euclidean division avoiding zero divisors *(with quotient and remainder).*	Yes	Yes	No
÷	Commutative avoiding zero divisors: *For every a, b, a ÷ b = b ÷ a.*	No	No	No
÷	Associative avoiding zero-divisors: *For every a, b, c, (a ÷ b) ÷c = a ÷ (b ÷ c).*	No	No	No

Postscript

Here we present solutions to some of the challenges laid down in this chapter.

> Find other examples of a procedure that can be applied to every pair of natural numbers and give a natural number as outcome.

We can 'mix' addition and multiplication in order to obtain another operation. For example, we can write a ☯ $b = (a + b) \times (a + b)$, or a ☯ $b = a \times b + a$, and so on. There are an infinite number of combinations possible. These examples may be of little interest for school mathematics, but the following examples of operations are more relevant.

In year 5 the first examples of raising to a power may be introduced. The National Curriculum (see Department for Education and Employment 1999) presents explicitly the cases one squared, two squared,... and writes 1^2, 2^2,.... These numbers can be related to the area of a square (year 4). Moreover, children learn 3D shapes very early and the volume of these geometrical shapes asks for numbers such as 1^3, 2^3,.... In this way we can think of these examples as instances of a binary operation 'raising to the power', different from addition and multiplication, also having both closure and uniqueness properties. This operation is neither commutative ($1^2 = 1$ and $2^1 = 2$) nor associative since $(2^3)^2 = 8^2 = 64$ and $2^{(3^2)} = 2^9 = 256$.

Some other simple examples are linked with the order. We can consider min and max operations: the former returns the smallest of two numbers and the latter the greatest of two numbers. For example, **min**(5,3) = **min**(3,5) = 3; **max**(7,4) = **max**(4,7) = 7. These two new operations have both the closure and the uniqueness properties; moreover, they are commutative and associative. Furthermore, **max** has 0 as neutral element, **min** has 0 as absorbent element.

In middle school children will study **gcd** and **lcm**, two other binary operations both commutative and associative.

> Could there be two different smallest elements of a non-empty subset of *N*?

The answer is no. A simple argument may convince us that if a and b are different and both smallest elements of a non-empty set *A* included in *N*, then we have a contradiction, since we would have $a < b$ and $b < a$ at the same time, but by the comparison property this conclusion is impossible. The same argument applies also to the case of 'greatest'.

Further reading

MacLane, S. (1986) *Mathematics: Form and Function*. Berlin: Springer.

For those who enjoy a cultural challenge, this book is a sort of conceptual encyclopaedia in which mathematics is treated 'from its inside', with a very great number of connections among mathematical topics. The resulting picture is fascinating.

References

Department for Education and Employment (1999) *The National Numeracy Strategy: Framework for Teaching Mathematics from Reception to Year 6*. Sudbury: Department for Education and Employment.

MacLane, S. (1986) *Mathematics: Form and Function*. Berlin: Springer.

Mangione, C. and Bozzi, S. (1993) *Storia della logica: da Boole ai nostri giorni*. Milan: Garzanti.

Appendix

Division by zero and one: a more mathematical explanation

Carlo Marchini and Paola Vighi

In this final part of the book we discuss zero, one and division. What follows may appear to be fairly technical, but we have endeavoured to make it as accessible as possible by including numerous examples to illustrate key points. Although it is not crucial that you, as a primary teacher, have a full understanding of the complexities of these topics, our view is that knowledge of more advanced mathematics is never wasted in the primary school classroom.

Some might describe division by 0 and 1 as a sort of mathematical nightmare, a ghost often found haunting high schools and universities. We can consider this ghost as a typical misconception which, as you will see, might well have arisen in primary school.

The Primary Framework document (Department for Education and Skills, 2006) refers to division across the primary years as set out in Table A.1. If you stop to think about it, some of the words related to division are quite tricky. For example, one does not immediately discern the presence of 2 in 'half' or the presence of 4 in 'quarter'. It is also interesting to note that the specific words 'dividend' and 'divisor' are not present in DfES vocabulary.

The idea of 'sharing equally' looses its meaning if the divisor is greater than the dividend (e.g. $4 \div 6 = \square$) but also if it is 1 (e.g. $4 \div 1 = \square$), therefore the early encounters with division could result in misunderstanding.

There is another problem connected with division. In Chapter 7 we pointed out that it was not always possible to use division within the natural number system. If we ask 'what is the result of $86 \div 25$?' it may not be clear what answer is required: is it 3.44 or 3 with remainder 11, or simply 3? And what is the result of $85 \div 26$? Is it 3 with remainder 7 or 3 or 3.2692307692307692307692307…, with an infinite number of decimal digits?

Division with a remainder, which is also called *Euclidean division*, can be performed for every pair of natural numbers (dividend, divisor), with the proviso that divisor is different from 0, and its result is a pair of natural numbers (quotient, remainder). These pairs (dividend, divisor) and (quotient, remainder) are ordered, since if the numbers are reversed the results are different: the division of 8 by 3 is different from the division of 3 by 8 (division is not commutative). In what follows we consider the 'result' of a (Euclidean) division of

Table A.1 Division across the primary years as set out in the Primary Framework (Department for Education and Skills, 2006)

Reception	'Share objects into equal groups and count how many in each group' (p. 70)
Year 1	'Solve practical problems that involve ... sharing into equal groups' (p. 72)
Year 2	'Represent ... sharing and repeated subtraction (grouping) as division; use practical and informal written methods and related vocabulary to support ... division, including calculations with remainders.' (p. 74)
Year 3	'Use practical and informal written methods to ... divide two-digit numbers (e.g. ... 50 ÷ 4); round remainders up or down, depending on the context. Understand that division is the inverse of multiplication and vice versa; use this to derive and record related multiplication and division number sentences.' (p. 76)
Year 4	'Multiply and divide numbers to 1000 by 10 and then 100 (whole-number answers), understanding the effect; relate to scaling up or down. Develop and use written methods to record, support and explain ... division of two-digit numbers by a one-digit number, including division with remainders (e.g. ... 98 ÷ 6)' (p. 78)
Year 5	'Use understanding of place value to ... divide whole numbers and decimals by 10, 100 or 1000. Refine and use efficient written methods to ... divide ... HTU ÷ U. Find fractions using division (e.g. 1/100 of 5 kg)' (p. 80)
Year 6	'Calculate mentally with integers and decimals ... TU ÷ U ... U.t ÷ U Use efficient written methods to ... divide integers and decimals by a one-digit integer. Relate fractions to multiplication and division (e.g. $6 \div 2 = \frac{1}{2}$ of $6 = 6 \times \frac{1}{2}$; express a quotient as a fraction or decimal (e.g. $67 \div 5 = 13.4$ or $13\frac{2}{5}$' (p. 82)

an ordered pair (quotient, remainder). In Italian and Israeli schools the customary notation $8 \div 3 = 2$ (2) or $8 \div 3 = 2$ (r. 2), in which it is clear that the two instances of 2 have different meanings, is used.

The fundamental property of this kind of division is as follows. Consider $86 \div 25 = 3$ (r. 11). The dividend 86 is equal to the sum of the divisor 25 multiplied by the quotient 3, added to remainder 11, that is, $86 = 25 \times 3 + 11$. Note that the remainder satisfies the condition $0 \leq 11 < 25$. This chain of inequalities ensures the uniqueness of the quotient and remainder, given the dividend 86 and the divisor 25. To record this in a more abstract mathematical manner, we can say that, in a (Euclidean) division – where the divisor is never 0 – the result is a (unique) ordered pair of natural numbers, given by the quotient and the remainder. These two numbers are such that the product of the divisor and the quotient added to the remainder is equal to the dividend and the remainder is greater than, or equal, to 0 and less than the divisor. In short,

dividend ÷ divisor = quotient (remainder)

under the following conditions:

- divisor different from 0;
- dividend = divisor × quotient + remainder;
- $0 \leq$ remainder $<$ divisor.

Thus, given the natural numbers 9 and 4, where 9 is the dividend and 4 is the divisor, we can compute $9 \div 4$ (since 4 is different from 0) resulting in a quotient 2 remainder 1, $9 \div 4 = 2$ (r. 1). The main property of division is satisfied, that is $9 = 4 \times 2 + 1$. In other words, the dividend (9) is equal to the sum of: the product of divisor (4) and the quotient (2) plus the remainder (1) and the remainder (1) is greater than or equal to 0 and less than the divisor (4). Note that we also can write $9 = 2 \times 4 + 1$, thanks to commutative property of multiplication and $0 \leq 1 < 2$. Therefore we can also state that $9 \div 2 = 4$ (r. 1).

Warning: $11 \div 4 = 2$ (r. 3) is equivalent to $11 = 4 \times 2 + 3$ but, contrary to expectations given the above, $11 \div 2 = 5$ (r. 1) as it is not the case that $3 < 2$!

Challenge

What is the result of a (Euclidean) division in which the divisor and the dividend are equal?

This is not a trick question and is, in fact, quite simple: consider sharing 9 marbles equally among 9 children, for example, or sharing 21 sweets equally in a class of 21 pupils. The answer is one each, in both cases. But this answer hides the fact that we are in the context of Euclidean division with remainder. Of course the remainder is 0, nevertheless this remainder must be mentioned, and therefore the answer is: the result of a division in which divisor and dividend are equal is 1 with remainder 0. There is the temptation – to which we often fall prey – to give only the quotient as the result of such a division problem, since 0 'is nothing' and when the remainder is 0 there is nothing to say! But in this case 1 and 0 are strictly tied. And remember that the dividend must be equal to a sum, of a suitable product together with a remainder. Therefore answering 1 to the challenge is not a complete answer as only the product – divisor × quotient – is presented, and thus we have one addend of the addition – whose sum is the dividend – but not the other.

Challenge

Can you find a division problem in which 0 is the quotient and 1 is the remainder?

This question points out one of the main misconceptions regarding division: that division with natural numbers is possible only when the dividend is greater than the divisor. This is a typical misconception almost certainly originating from the fact that in school teaching and in textbooks most examples of division are with dividend greater than divisor, since the aim is to teach the division algorithm. For example, you almost certainly would not see $3 \div 7 = \square$ in such a book. The model of division of cakes is sometimes misleading: the slice is not a cake; moreover, it is smaller than the whole cake. Furthermore, a cake is something of a continuous thing in that, in theory, it can be cut into any number of slices of varying size. In contrast, Euclidean division works for natural numbers (a discrete set) and the 'result' is the (ordered) pair: quotient and remainder. The only requirement for a (Euclidean) division is the fact that the divisor must be different from 0, and no other condition is necessary. Thus, for example, $1 \div 132 = 0$ (r. 1), or $1 \div 3 = 0$ (r. 1). But $1 \div 1 = 1$ (r. 0) with a divisor of 1 is a unique case in which the dividend 1 does not give quotient 0 and remainder 1. In other words, the answer to the above challenge is simple. Put more abstractly, the dividend is the sum of the product of divisor times the quotient (0) added to the remainder (1). Since a factor of the product is 0, the product itself is 0 and the sum is 1, since the remainder is 1. But the condition for the remainder is $0 \leq 1 <$ divisor, hence the divisor is any natural number greater than 1. Note that $3 \div 7 = 0$ (r. 3) and that in every case in which the dividend is less than the divisor the quotient is 0 and the remainder equals the dividend.

Challenge

Which division problems give the quotient 0?

Which division problems give a remainder which is equal to the dividend?

Which division problems give a quotient of 0, remainder 0?

This challenge is quite complex, but with patience we can unwind the skein. Take, for example, the number 18 as a dividend in a division with quotient 0. For the main property of division we have 18 = divisor × 0 + remainder. Now it is evident that whatever the divisor – given that we want the quotient to be 0 – the product divisor × 0 = 0, therefore 18 = remainder. But in every case, the divisor can never be 0 and 0 must be less than, or equal to, the remainder (18) which, in turn, is less than the divisor. Therefore we get 18 < divisor. In the above example, the divisor happened to be 18 but it could have been any other natural number such as 23, or 19, or 156. We can state that when dividing, if the quotient is 0, the dividend is equal to the remainder and the dividend is less than the divisor. Conversely, if dividend and remainder are equal, then divisor is greater than the dividend and, therefore the quotient is zero: 18 = divisor × quotient + 18, therefore divisor × quotient = 0. But divisor is different from 0 and hence quotient must be 0.

Therefore the conditions: dividend < divisor, quotient 0, dividend = remainder are equivalent.

Now the answer to the third question is easy to find. Using the main property of (Euclidean) division, the given data are enough to assert that the dividend is 0 since divisor $\times$ 0 + 0 = 0. We have only to choose the divisor, but, as every natural number is different from 0 is a suitable divisor: e.g. 0 ÷ 5 = 0 (r. 0); 0 ÷ 1 = 0 (r. 0); 0 ÷ 1234 = 0 (r. 0).

Challenge

Which divisor can give 0 as a remainder, whatever the dividend?

When the remainder is 0, the dividend is divisible by the divisor, since dividend = divisor $\times$ quotient + 0. Let us experiment by having a dividend 0; hence, for the above challenge, the quotient would be 0 and remainder would also be 0, for every divisor. Now let us try with a dividend 1, thus 1 = divisor $\times$ quotient + 0. But the product of two natural numbers is 1 if and only if both factors are 1, which means that the divisor is 1 and the quotient is 1: in other words, 1 ÷ 1 = 1 (r. 0). Is it the case that a divisor of 1 fulfils the condition of the challenge? By way of response, consider the following:

- We know when the remainder is 0, whatever the dividend, 0 ≤ remainder < divisor.
- We also know that if the divisor is 1, then 0 ≤ remainder <1.
- Using the fact that the set of natural numbers is discrete, we can conclude that the remainder is 0.
- Therefore, in this case dividend and quotient are equal: for example, 8 ÷ 1 = 8 (r. 0), 13 ÷ 1 = 13 (r. 0), 0 ÷ 1 = (r. 0).

We can also deduce that division by 1 gives a dividend which is equal to the quotient and a remainder of 0.

Challenge

Why must a divisor be different from 0?

In the light of the above, consider the Table A.2, bearing in mind that we have stressed that 0 can never be a divisor. Notice how:

- In the first three rows a divisor greater than the dividend gives a quotient 0 and a remainder equal to the divisor. The table considers just two examples of the infinite number in which the divisor is greater than 36.
- In the fourth row, if dividend and divisor are equal the quotient is 1 and the remainder 0.
- From the fifth row on, the dividend is greater than divisor: the remainder oscillates, while the quotient grows 'regularly' until the last four rows.

- From the fifth row onwards there are several occasions – in total 36 – where, despite variations in the divisor, the same quotient (other than 0) results. More specifically, had we included all the divisors from 38 down to 1 we would have had 18 quotients of 1, 6 of 2, 3 of 3 and 2 of 4.
- If you cast your eye down the quotient and remainder columns from the fifth row onwards you will see that if the quotient is 1 the remainders increase by 1, if the quotient is 2 they increase by 2, and so on.
- In the last four lines the remainder is 0 on the basis of the divisibility condition, since all the natural numbers 1, 2, 3, 4, are divisors of 36.
- In the last line where the divisor is 1, the quotient equals the dividend and the remainder is 0.

Table A.2 Some division calculations with 36 as the dividend

row	dividend	divisor	quotient	remainder
1	36	...	0	36
2	36	38	0	36
3	36	37	0	36
4	36	36	1	0
5	36	35	1	1
6	36	34	1	2
7	36	33	1	3
8	36	...	...	...
9	36	19	1	17
10	36	18	2	0
11	36	17	2	2
12	36	16	2	4
13	36	15	2	6
14	36	14	2	8
15	36	13	2	10
16	36	12	3	0
17	36	11	3	3
18	36	10	3	6
19	36	9	4	0
20	36	8	4	4
21	36	7	5	1
22	36	6	6	0
23	36	5	7	1
24	36	4	9	0
25	36	3	12	0
26	36	2	18	0
27	36	1	36	0

Note that the table does not have a row in which the divisor is 0 but we can imagine what might happen with such an (impossible) divisor.

- The regularity in the table could suggest that as long as the divisor decreases, the quotient increases. Thus we might deduce that if we divide 36 by 0 the quotient would be greater than 36 (*continuity argument*).
- Or we could consider what would happen if the remainder were greater than or equal to 0 and less than the divisor. If the divisor were 0, the remainder would have to be a natural number greater than or equal to 0 and, at the same time, less than 0. But there is no natural number less than 0, since 0 is the smallest natural number (*discreteness argument*).

Here lies the conundrum we face if we want to divide by 0: from one point of view we must have a quotient greater than the dividend, and from the other we cannot have a remainder! In other words, the concepts of continuity and discreteness are incompatible.

Remembering the trend of a growing quotient as the divisor decreases is useful when dividing by decimals or by fractions less than 1, as in such cases the intuitive view (see Chapter 3) that division makes smaller might be violated.

The peculiarities of division by 0 and 1 are often not considered or, as we discovered in some textbooks, inadequately discussed. Indeed, the question of having a divisor equal to 0 is often passed over in silence and yet it is important later in school life when the fractions are used and much later on, when the incremental ratio is used in derivative calculus. It is tempting to provide you with the following conclusion, 'Division by 0 is impossible. From now on, please remember: you can never divide by 0.' Such a generalisation, however, is not strictly accurate and is, indeed, contrary to the spirit of this book where we countenance against making statements which do not universally apply. It is more accurate to say: 'When you are working with *real numbers* (natural, integer, rational and irrational numbers) please refrain from division by 0. Dividing by 0 is not allowed when you are working with these numbers.' Thus when the more mathematically minded members of your class encounter concepts such as infinitesimals, they will be prepared rather than having to confront a previously held misconception regarding division and zero.

References

Department for Education and Skills (2006) *Primary Framework for Literacy and Mathematics*. London: Department for Education and Skills.

Index

Note: numbers are entered under name (e.g. 'ten'). The letter 'f' refers to a figure; 'p' to a picture or photograph; 't' to a table.